My Bout with Multiple Myeloma

written by a 60 year old, formerly healthy,
male Christian

Dennis R. Dinger

My Bout with Multiple Myeloma

Copyright © 2009
Dennis R. Dinger

ISBN 978-0-557-82605-6

Published by: C B Dinger

*to all multiple myeloma patients
and their families*

Contents

Acknowledgments

Primary thanks go to the Lord for supporting my family and me through this whole process, for putting me in the right place at the right time, for preventing me from making bad decisions regarding cancer treatments, for making my body cooperate well with all medications and procedures, and for providing His love and grace to us.

Next, thanks go to my beloved wife, Christine, who has been my primary care giver and supporter throughout this whole process. She has helped in innumerable ways to make life easier and more comfortable as I endured all procedures and treatments. I love you! Thanks!

I want to thank my family doctor, Dr. W. Douglas Gentry, for persisting with the lab tests long enough to diagnose my cancer and for referring me to an outstanding oncologist. I want to thank my primary oncologist, Dr. Mark O'Rourke, and the other doctors, nurses, and staff at the Cancer Centers of the Carolinas Seneca Office for all they have done for me. I especially want to thank the chemo nursing staff in Seneca who spent many hours providing care during my many chemotherapy sessions. Thanks also go to my stem cell transplant doctor, Dr. Gary Spitzer, and the other doctors, nurses, and staff at the Cancer Centers of the Carolinas Eastside Office in Greenville and at the Diagnostics and Therapeutics Department at St. Francis Hospital in Greenville. I especially thank the transplant nurses at the Eastside Office who spent many hours providing care during the transplant and high dose procedures. The professionalism, expertise, and care for me and my family exhibited by all of these people has been truly outstanding. I am blessed to have been under your care. Thank you all.

My son, Matthew, drove and accompanied me to many appointments, took care of me at home, and functioned as care giver during many days of my first stay in a hotel in Greenville. My nephew,

Joel Funk, functioned as chauffeur and care giver during many days of my second stay in Greenville. My daughter, Rachel, my son, Daniel, and my sister-in-law, Betsi, also helped during my many appointments in Greenville. My other sons, Jon and Joe, helped to make my life easier at home. Thank you all!

Friends who took time away from their busy schedules to drive me to appointments in Seneca and Greenville include Beth Taylor, Flo Jachens, and Elise Clark. It would not have been possible for me to make all of the appointments without your help. Thanks.

Special thanks to Trev and Angela Alberts and Flo and Darryl Jachens for their support, words of encouragement, and prayers. Thanks also to Keith and Diane Van Ryn, and the saints at the Summerville Bible Fellowship. Thank you and God bless you all!

Thanks to the many students, faculty, staff, parents, and friends at Oconee Christian Academy in Seneca for your cards, phone calls, words of encouragement, and prayers. Thanks to my many friends around the country for your words of encouragement, cards, notes, and prayers. I am blessed to have your support and to be in your prayers.

The Lord provided me with excellent medical treatment as well as wonderful moral and spiritual support throughout this whole cancer treatment process. To all of you who participated in any way, I owe you my gratitude. I know words don't and can't adequately express my thanks, but they will have to do. Thank you all! God bless you!

Preface

One of the first notes I received from a friend after announcing that I had been diagnosed with cancer was the suggestion that I keep a diary so I could write "my story." My friend suggested that since there aren't many such books available, my experience could be quite helpful to others.

I have not really been keeping a diary, but I decided I would attempt a book. Same thing. Since I started writing in the middle of the chemotherapy sessions — I have recorded events both as I remembered and as they happened.

My Faith

As a Christian, the Lord has supported, strengthened, and guided my wife, my family, and me throughout the cancer treatment process. He has provided me with an enormous support group of Christian friends from all around the country who are praying for us daily. And I have even received well wishes and prayers from business acquaintances in the far reaches of the world.

Most decisions during treatment were relatively minor. I just needed to agree with the recommendations of the doctors and go with the flow without having to really decide much of anything!

Sometimes, decisions were very important! Especially beneficial to me were those occasions when the Lord stepped in to make sure I didn't mess things up by making my own (bad) decisions. Those times reminded me of the verses in Proverbs and Jeremiah that say, **"There is a way which seemeth right unto a man, but the end thereof are the ways of death."**, (Pro 14:12) and, **"O LORD, I know that the way of** man is not in himself: it is not in man that walketh to direct his steps." (Jer 10:23)

When the Lord took decisions out of my hands, **I could not** make the wrong decision. Hindsight says I **would have** made the wrong decision! In some cases, I had no choice: I **didn't** want to be admitted to the hospital over Labor Day, 2008 — but I didn't have a choice! They admitted me! I'm sure my wife approved that decision — but she really had no choice. I **did** want my crushed vertebra fixed — but I didn't have a choice — my blood was too thin that morning and the neurosurgeon wouldn't perform the procedure to expand and strengthen one of them. I thank the Lord for those days!

I know non-Christians who have gone through intense cancer treatment. I know non-Christians who had terminal cancer, whose disease was too far along to give any treatments at all. I have heard them speak. I have read their words. Without the Lord on their side, their lives, their experiences, and their final words ring hollow.

I cannot imagine anyone going through this type of ordeal without the Lord's help. The path traveled during chemotherapy, treatment, and recovery is tough, to be sure.

The outlook taken on life and life-expectancy after being diagnosed with cancer also depends upon one's relationship with the Lord. I am a Christian. The Lord has been with me throughout the diagnosis, treatment, and recovery process. I praise and thank Him for His support!

Your Faith

I encourage any of you who are reading this, who do not presently know the Lord, to inquire about Him. Ask a knowledgeable Christian friend to explain the Gospel and the truth of God's salvation to you. He offers great spiritual and emotional strength and support, and He can offer great physical strength, support, and healing as well. Or — contact me. I'll explain it to you in great detail.

The Future

Upon hearing of my cancer, another of my friends told me that the question I needed to consider was, "What does the Lord want me to do with the rest of my life – post cancer?" That answer depends on His goals for me as an individual, and I trust He will make those goals clear as I progress through and beyond this cancer. That question is constantly in my prayers.

Cancer affects every individual differently. Some succumb to it. Some get past it and continue almost as if nothing had happened. Some get past the cancer, but must deal with permanent debilitating side effects caused by the cancer or its treatments. It varies. In my case, this cancer has caused deterioration to my bones and spinal column. I have lost 3 inches in height which may never return, although some of the "extensive" deterioration to my spinal column **may be** (????) reversible with appropriate exercise and therapy. The full extent of the peripheral neuropathy in my hands and feet may or may not be reversible. My heart problem appears to be reversible. Each individual can expect similar but unique problems of their own.

Anyone who must face cancer and chemo treatments needs good support. Constant encouragement leads to an optimistic point-of-view. Without the help of others, it is easy to be pessimistic, to become depressed, and to simply give up. On the days when my white blood count was close to zero and I had no energy whatsoever, I could see how easy it would be to say, "This is not worth it! I quit!" But giving up was not an option. Only in hindsight did it even occur to me that quitting could have been an option.

I continue to look optimistically to the future, to deal with the cancer, and to live with its side effects. I hope this book will help others learn from my experiences so they can go through their own cancer, treatment, and recovery processes in good spirits.

God bless each of you who are reading this because you are experiencing something similar. God bless each family member who has to care for a loved one who is dealing with a cancer. God bless each one who is reading this just to hear my story. May this written work be a blessing to you all.

                              Dennis R. Dinger
                              10 July 2009

Section 1

Background

1

My Background

I am a 60 year old white male who is a Christian, a husband, a father of five, an engineer, an Emeritus Professor of Ceramic and Materials Engineering at Clemson University, a consultant, and a high school science and calculus teacher. One week after my 60th birthday, I was diagnosed with multiple myeloma.

My health has been good most of my life. I have almost always been overweight and my blood pressure has been borderline-high for many years. I only started taking medications for high blood pressure a few weeks before being diagnosed with cancer.

I have rarely seen our family doctor during our twenty year tenure in Clemson, SC. In fact, when I went to see him in the spring of 2008, when I was beginning to hurt from the effects of this cancer, it had been so long since my last office visit that I had to fill out a 'new patient' history form. One of his nurses, who knows me, asked in disbelief if I hadn't ever seen him before. Turns out that my files were so old, they had been moved to their archives.

It must have been close to ten years since I had seen the family doctor for anything. My last visit was probably for toenail fungus, an ear ache, or something else equally critical. Throughout my adult life, I subscribed to the idea that I would only go to a doctor when I was sick. Scheduling an office visit for a headache, a slight fever, or some other minor ailment, only to be told, "Go home, take some aspirin, and see how you feel tomorrow. That will be 50 bucks, please!" has never been my idea of money well spent. So I only went when I thought I needed the doctor's help — which wasn't very often.

Even during the spring of 2008, when I was starting to hurt from the cancer, I self-diagnosed and self-treated as long as possible. The teacher in the next classroom at school said it was in my genes, "Men are all like that!" My wife wanted me to go to the doctor sooner, too, but I told her the doctor would only tell me to do what I was already doing **and** to take some aspirin. In fact, when I did finally visit the doctor, that was exactly what he said. "Take it easy, don't do things that will aggravate the pain, and take an anti-inflammatory" (he then named a specific over-the-counter pain killer I should take — instead of aspirin.)

I was correct in my initial self-diagnosis and self-treatment. Both the doctor and I thought I had a bruised rib and an inflammation of the rib cage. After the second visit, when the prescription anti-inflammatory did not solve the problem, it began to appear that I had something more serious and more elusive.

About that same time, my sleeping began to be interrupted by frequent bathroom breaks about two hours apart. Four such cycles made up an 8 hour night. This, too, was totally foreign. I have always been able to sleep well. Usually within about 5 seconds of my head hitting the pillow, I am out, sound asleep, and I usually sleep all night without bathroom breaks. But all that had changed!

Minor pains here and there around my body were routine. I had no major pains, though. Occasionally, I have tweaked my back, which would set me up for a few weeks of nasty lower back pains. Those episodes, however, were few and far between. In most cases, I could do heavy work without any problem. If and when I popped my back, it was usually for something trivial, like bending over to pick up a piece of paper or to tie my shoelaces.

I have had carpal tunnel problems in my hands and fingers for years. Those problems are worst when I do heavy work with my hands. If I decide to build something major, I set myself up for tingly, hurting hands and fingers. The summer I laid pavers to create a brick pathway from our driveway to the back of the house, severe carpal tunnel problems appeared. Picking up two or three pavers at a time in each hand required lots of force by the fingers and that caused my hands to hurt most nights during the project. When the project was finished and the strenuous demands on my hands stopped, the pain and tingly sensations went away.

Over the years, I have spent many hours each day typing at a computer. I have been using ergonomic keyboards ever since they were invented, and that has allowed me to type with my hands and fingers in relaxed positions. Typing, therefore, has never caused any carpal tunnel problems. I do lots of work at the computer, so if carpal tunnel problems were derived solely from keyboard work, I should have had continuous, severe carpal tunnel problems starting years ago. Unless I'm doing heavy, physical labor, however, I don't.

2

The Medical Team:
The Doctors, Nurses, & Coordinators

The Medical Team

I am blessed that my diagnosis occurred in 2008. Cancer medicine and treatment procedures made major strides during the most recent two decades.

When we moved to Clemson, SC, 20 years ago, we were fortunate to find a good Christian family doctor whom we trust for both medical and spiritual advice. There are so many doctors, scientists, and teachers around today who are evolutionists and not Christians. Biology is taught in most schools from an evolutionary point-of-view. After moving here, we were fortunate to get a good recommendation which directed us to this Christian family physician.

Then, when it came time to refer me to an oncologist, the family doctor sent me to the oncologist he said he would want to be treated by if he were in my shoes. Turns out that the doctor he sent me to is the same doctor who treated my mother about nine years ago when she had cancer. So from the day I was referred to him, I had great confidence in his knowledge, skills, and abilities as a physician.

My whole team of doctors and nurses have been excellent. My primary oncologist works in both the Greenville and Seneca Offices of the Cancer Centers of the Carolinas. He shares an office complex in the Greenville clinic with the Stem Cell Transplant Director, and he shares an office complex in the Seneca clinic with other excellent doctors. As I have interacted with them, I learned that they constitute a wonderful medical team.

The medical team has been totally optimistic about my progress throughout this whole treatment process. They all constantly referred to me as a "young and healthy" man who was capable of handling the aggressive chemotherapy regimen. My primary oncologist explained that life expectancy after diagnosis, following the aggressive treatment he prescribed, was "10$^+$ years and growing." After this treatment regimen, some patients have been in total remission for more than 10 years. They are monitored annually and with no new signs of the cancer, those people just keep pushing the life expectancy envelope number higher and higher.

During my chemotherapy days, when I looked in the mirror, I did not see a "young and healthy" individual looking back at me and I certainly did not feel healthy. But whenever the medical staff described me as "young and healthy," I know they were comparing me to other patients who are much older than me and/or who are much further along in the development of their cancers. Relatively speaking, I handled the treatments remarkably well.

My viewpoint has been based on only one experience with cancer — my own. One experience is very limited data on which to base any evaluation, so I learned to trust the medical staff when they said I was doing really well!

If I had been 80 or so, and feeble, they would not have described me as "young and healthy" and they would not have subjected me to the aggressive chemo treatment. I would have received a less aggressive, less stressful treatment. In fact, just before beginning cycle four of chemotherapy, a transplant coordinator reminded me that some people simply cannot take four cycles of my particular chemotherapy. Depending upon how one tolerates initial treatment cycles, team doctors decide how best to proceed and how many cycles of chemo the patient can handle. I did well so I received the full treatment.

Younger people are generally good candidates for the same aggressive treatment — providing they are sufficiently healthy to handle it. For multiple myeloma, the aggressive treatment I received was the one with the most promise for full remission and longevity of life post-cancer.

Multiple Myeloma Treatments & "Cures"

The Transplant Director shared with me a presentation by his research colleague who suggested that when a multiple myeloma survivor has been in complete remission for more than 10 years, doctors should consider telling them that they have been "cured." Up to this point in time, however, the "C"-word has not been used for multiple myeloma patients.

The "C"-word "cure" is a word used very sparingly by the oncology community. Many cancers remain incurable. Everything in the literature for multiple myeloma describes it as incurable. They still don't know what causes it, and they don't know how to cure it. Even though some patients have been in remission for more than 10 years, the day has not yet arrived in which they are willing to use the "C"-word with multiple myeloma. There is certainly hope on the horizon that they may start using it soon.

Current Research Literature & Treatment Regimens

When reading literature on any type of cancer, one must search for the most recent papers, presentations, and articles to find the current, best information on treatment processes and cures. In my case, there were lots of several-year-old articles on the internet which were optimistic, but which didn't show particularly good projected outcomes. The Stem Cell Transplant Director, however, pointed me to two recent slide shows presented by a research colleague. Those presentations contained the most up-to-date, best, and most optimistic information about treatments and successes for multiple myeloma.

Regardless what doctors may explain concerning your cancer, you should search out the most up-do-date, relevant articles to read for yourself. If necessary, request copies or references. It is in both the doctors' and the patients' best interests that everyone be up-to-date and up-to-speed on the details of your particular cancer and on the nature and course of treatments available for it.

Seek second, third, and fourth opinions as you feel necessary. Depending upon the urgency, time may be available to explore alternative and/or homeopathic treatments. Such treatments are not necessarily easy to find, but they are out there. Family and friends may help point you in appropriate directions.

In my case, it was fairly urgent that treatment start as soon as possible. At the time of my diagnosis, I was in Stage III of III. Right from the start, I felt the Lord was in control of my situation — which included my treatment — so I felt no need to search beyond the treatment recommended by my medical team. I was comfortable that the Lord had arranged the best treatments for me from the best doctors and I searched no further.

I know others, however, who feel totally different. They want multiple opinions and/or homeopathic treatments. There is no one absolutely correct way to treat cancer. Whatever you feel you need to do is a GOOD way to proceed! There is enough stress in the cancer diagnosis and treatment processes without getting overly stressed about the intricacies of research results. Make your decisions based on "sufficient" information. You (the patient) must determine how much information constitutes "enough." When you have gathered **enough** information, decide how you want to be treated. If you don't have **enough** information, keep searching until you do. Then, choose a treatment path!

The new patient usually knows the least about his/her cancer, but ultimately, the patient is the one who must decide how to treat the cancer. Either support the doctors' recommendations, or search for other doctors — until such time as you find one whose recommendations you support. It is your choice. Find a good medical team and then trust them that they know their business. My medical team knows the cancer treatment business and I trust their opinions. I still am ultimately responsible to make the decisions, however! I am constantly reminded of this because I must constantly sign consent forms to allow any and all treatments.

3

Support Group

Since the time I announced my cancer to family and friends, I learned that I have an extensive Christian support group from all around the country and the world who have been praying for me on a daily basis. The prayers, concerns, encouragements, cards, well wishes, phone calls, e-mails, etc., from this group have been wonderfully uplifting. It brings tears to my eyes every time I think about these friends and acquaintances who are holding my family and me up to the Lord.

As I indicated earlier, this support has come in all different forms. Some call and pray with me. Some e-mail. Some send cards. Some drop by to visit. Some want short, frequent e-mail updates of my condition. Some want long detailed updates. Some poke fun that I don't know how to write short updates. One friend told me I could write another "War and Peace" volume and that would be fine. Regardless, I'm sure they are all praying for me and the family, which is the amazing part. I don't deserve any of this, yet I am receiving it anyway!

I pray that each of you who must walk a similar path through cancer treatments to recovery have an equally large support group behind your every step. It makes all the difference in the world!

4

Finances – Insurance

Some History

I have always said it was easier **to talk** about living by faith than **to actually live** by faith. For years, when I was employed in academia, we had health insurance and we counted on a biweekly pay check. We may have **said** that the Lord was providing for our lives (and He was), but we really relied on the knowledge that I had a steady income. We lived a very comfortable life. When I retired from the university, that all began to change.

I retired to take a consulting opportunity in the Far East. This required four month-long trips each year over a period of several years and long stays in Semarang, Indonesia. As a self-employed consultant, it also required that I find my own health insurance.

Initially, it appeared that the consulting in Indonesia would continue until my retirement. I look back on those days in Semarang with fond memories. But as that consulting opportunity began to wind down, I was forced to find income elsewhere.

I did not advertise in the US while working in the Far East, but when that consultancy ended, I began to do so. Our income decreased considerably then and the retirement monies from my academic jobs were used up quickly to maintain our health coverage.

Long ago, my wife's arthritis doctor told us to make sure we kept good health insurance because the next prescription medication my wife would need to take (after her current medication lost its effectiveness) was "really expensive." We were covered — but I needed to search for less expensive insurance. My wife's medicine was beginning to lose its effectiveness, so the new, expensive prescription was just around the corner.

Individual plans were one option. When we applied for **individual** coverage, we learned that her arthritis was considered a *pre-existing condition* which would not be covered. She was also routinely denied coverage because of her arthritis meds. No company would cover her with an individual health plan. The children and I qualified for policies, but my wife did not. We needed **group** coverage — because the law requires insurance companies to insure **everyone** who applies within a group. The trick was to be in a group.

We were covered during my academic years by the universities' group health care plans. When I left academia, to continue group coverage the employees of my consulting company (my wife and I) became a **group** for insurance purposes. We formed the smallest possible group (a group of 2) which made it the most expensive group size. But we were a group — and the whole family (including my wife) was covered.

That coverage was really expensive. Effectively, we had two policies: one policy covered me, and the other covered my wife and the children. Paying for two policies rather than one helped to make the insurance really expensive. To reduce premiums as much as possible, we had taken a plan with a $10,000 yearly deductible — which basically meant everything but a major hospital stay would be covered by our nickel. But we were covered!

Even with that extremely large deductible, premiums continued to increase. When my retirement money began to run out and the consulting was insufficient to make ends meet, both my wife and I began to search for new jobs. At the time, we figured that one of us could work and get health care coverage while the other could home school our youngest son Joe to finish his high school education.

One day, I inquired at the local Christian schools in Pickens and Oconee Counties in South Carolina. I had made a mental list of local places to check for jobs, and that particular day, I remembered my list.

The school in Oconee County invited me right over to talk. Turns out they needed a science teacher. Everything I said I could teach — they needed!

Within a week of my inquiry, I had a job teaching six courses at OCA: algebra, calculus, earth science, physics, physical science, and biology. The job had no health benefits, but Joe could go to school there tuition-free. That meant my wife remained free to search for a job with health benefits and neither of us needed to home school Joe.

As a Christian school, OCA provided a steady income, but it was much smaller than we had ever had — and it did not provide health benefits. If I said, "Yes," to the job, we were definitely going to be relying on the Lord. I said, "Yes."

Primary Health Care Coverage

My wife, Chris, and I have always believed that we needed good health care coverage for our family: When one of us becomes sick, we should be covered. That is a good plan, but it can be difficult to achieve. Fortunately, a few days after Joe and I began traveling to Oconee Christian Academy, Chris was offered a food service job on the Clemson University campus. They offered health insurance to full-time workers, and Chris' new position was full-time. We were covered!

I am convinced that the Lord guided us into these jobs because He knows what He is doing and He understands our needs — even though we didn't understand what was happening at that time. His guidance and preparation only became obvious to us after the fact.

We each had questions about our job status and health insurance status: Why was I doing what appeared to be the Lord's work at OCA while Chris was not doing the Lord's work at the cafeteria? Why did I like my job? Why did Chris dislike hers? Why were both of us doing things that were inconsistent with our educations and our perceptions of the Lord's desire for our lives?

We still have questions to this day, but if we add the dollar amounts of the huge medical bills that were paid by Chris' health insurance to her wages, she is making more money than either of us ever brought in at any time during our married lives. We are covered! My cancer treatments are covered! Chris' arthritis treatments are covered!

Chris announced to her boss one day that she had just gotten a raise. The boss excitedly asked, "For how much?" When Chris responded with $75,000.00 for that month alone, the boss' jaw dropped. Chris explained that was about how much the health insurance paid for my first month's medical treatments. During the whole cancer treatment process, we will recoup all of the health care monies we ever paid into any and all health care plans throughout my whole industrial/academic career — and then some! All of this is from the health coverage provided by Chris' food service job.

We are covered and we are thrilled! We have the Lord on our side and He has guided us! He has seen fit to put us where we are, and fortunately, both of us have had enough patience to go along with His program even when we didn't see where He was leading or why.

Group policies through employers are limited to the terms of the particular health care plans arranged for their employees. Some are better than others. When changing jobs, the level of coverage provided with the new job, and by the new company's plan, needs to be a consideration — especially when health issues exist. Certainly, as you age, it is important to pay attention to the levels, quality, and cost of coverage provided by health insurance plans at any new job.

We have had continuous group coverage for our family ever since I graduated from college. Through the years, my wife has made the most use of our coverage. I used it very little — that is, until I turned 60 and my body self-destructed with this cancer. One simply does not know when one is going to need health care coverage.

General Costs

As I indicated above, my wife's health insurance is excellent — we couldn't ask for a better policy. But the plan doesn't cover absolutely everything. Until we reach a nominal out-of-pocket expense limit **each year**, the plan pays 85% and we pay the remaining 15%. And each time I visit the clinic, I pay a $30 co-pay. That doesn't seem like a lot of money, but daily co-pays for weeks of chemo, X-ray sessions, and other treatments add up quickly.

Costs for Medications

On our plan, we pay $15 for inexpensive meds, $30 for moderately expensive meds, and $45 for expensive ones. This, too, adds up quickly. The first day we met the oncologist, he gave me three or four sheets of prescriptions with multiple meds listed on each. I think there were over a dozen different meds to be filled that first day — each of which cost at least $15. That first visit to the pharmacy cost about $200.

Annual Out-of-Pocket Limit

Annually, we have a total out-of-pocket limit of several thousand dollars (initially, it was $3000.) It depends which, and how many, treatments are performed to determine how fast one will reach the out-of-pocket limit.

I was diagnosed on 3 July 2008, which means there were less than 6 months remaining in that calendar year. I reached my out-of-pocket limit by October in 2008. In 2009, I did not reach my out-of-pocket limit.

My wife and I hadn't planned for any of this. So the confluence of several phenomena (disability, no income, chemotherapy, medical bills, diagnosis procedures, treatments, etc.) hit us when we least expected it and were least prepared for it. Our insurance coverage helped me to obtain good cancer treatment and it helped us survive.

Supplemental Health Care Coverage

Another health care coverage we *stumbled* onto was a supplemental policy. Both Chris' and my mothers had supplemental policies that they bought when they were in their 60s. Chris' Mom had a nursing care policy; my Mom had a cancer policy; and each mom used their supplemental policy to its fullest extent.

In neither case do we know who initially inquired about the policy, why it was bought, or how the mothers acquired their supplemental policies.

Even though Oconee Christian Academy didn't offer any primary health care coverage, they brought in representatives of supplemental health care plans for our benefit. For a modest monthly fee, I jumped at the chance and arranged accident, sickness, cancer, and heart policies. My son's motorcycle accident was covered by the accident policy and my cancer was covered by the cancer and sickness (disability) policies. We benefitted from these plans — they helped us make ends meet especially during my chemotherapy treatments, disability, and time of zero income.

Some may consider supplemental insurance (... actually, all insurance) to fall into the category of gambling. None of us knows what tomorrow holds. We must therefore decide whether or not to purchase insurance "just in case."

The first engineer I worked with 40 years ago in industry used to brag that he kept all of his bankbooks in a shoe box. He claimed he never bought any type of insurance. If he, his wife, or his children became sick, he paid the bills himself. He put all of the money he would have spent on insurance straight into his savings accounts, so his accounts, which were numerous, were all full. That is one way to do it. And it certainly appeared to work for him.

To purchase or to not purchase a supplemental policy is an individual decision. In our case, we opted for supplemental policies for very specific problems consistent with our family's medical history. These were valuable additions to our insurance coverage.

Supplemental plans are very restricted in both nature and coverage. For example, our accident plan doesn't cover sickness or cancer issues — it specifically covers injuries due to accidents. It doesn't pay a huge amount for accidental injuries, but the amount paid helps meet medical obligations. Regarding the motorcycle accident, the main auto insurance coverage took a long time to work its way through the system — it dragged on for 1½ years. The supplemental policy's payout helped us meet medical bill obligations immediately following the accident.

As it turned out, there was a disability rider in my cancer policy that covered my first year's disability due to cancer. Most of my supplemental benefits came from the cancer policy. It paid various amounts for cancer meds, anesthetics, radiation therapy treatments, other procedures and treatments, etc., during all aspects of the cancer program.

The supplemental policies have been outstanding additions to our insurance portfolio. Hopefully, we will not need all of the policies we purchased. Having them available provides us the comfort of knowing that we have taken financial steps to provide for all possible contingencies.

Why Supplemental Coverage?

Why? If our primary health care plan is really great, why should we need or want supplemental coverage? The answer is fairly simple and it became especially obvious as I was going through chemotherapy.

Our primary coverage paid 85% of all charges until we met our annual out-of-pocket limit. After that, it paid 100%. Our part, 15%, doesn't seem like much, but 15% of lots of expensive procedures adds up quickly. The supplemental insurance monies typically arrived a month or two after each procedure was performed. By then, our primary insurance had paid its share and our obligations for the remaining 15% had come due. If I was diligent and prompt with the supplemental insurance paperwork, the bills for our share of all procedures arrived about the same time as supplemental payments.

Supplemental insurance is targeted very narrowly, so the policy only pays out in very specific cases. The vast majority of people who own such policies will never need them. So costs remain fairly low.

For those individuals who contract cancer, several factors come together to put the economic squeeze on the patient and their family. In my case, they were:
- total disability – i.e., no employment and zero social security payments for the first 5 months,
- good but not 100% primary insurance coverage, and
- constant co-pays and co-insurance requirements.

Medical Costs With & Without Insurance

During my chemo treatments, I learned about today's ridiculously high cost of medical treatments! I also learned that costs with a good health care plan are lower than costs with a poor health care plan.

Some of my medications came in subcutaneous injectable form. The pharmacy gave me pre-filled syringes and I simply needed to inject them at the appropriate time. The medications in those syringes were seriously expensive! If I were paying the full costs of medications, I wouldn't have been able to afford them.

During chemo weeks, one of the cancer drugs they gave me was administered as three shots over the course of several days. They injected this drug on days # -2, #1, and #4 of each chemo cycle. For each shot, $3010 was billed directly to my health care provider. I only needed to pay the $30 co-pay at the beginning of each chemo session to receive the injection.

Once the health insurance company paid their share and the bills for my share started to roll in, we were able to see what the primary insurance company was doing when billed for such drugs. In the case of that cancer drug, $3010 was billed for each dose, $1699 was "written off," and the primary health insurance company paid the remaining 43% ($1281). If I had no health insurance plan at all, or a poor one, I would have been responsible for the whole $3010 for each shot.

Why mention this? What is the point? With good health insurance coverage, the doctors and clinics have agreed to accept certain amounts for each particular treatment. These negotiated amounts are considerably lower than the billed amounts. Without good health insurance, the doctors and clinics expect payment-in-full for all billed services.

Our costs in each case were considerably less than they appeared simply because of the insurance company's pre-negotiated prices. We benefitted from that! It brought the part of the bill for which we were responsible down into a dollar range that could be covered by our limited resources.

We see similar types of numbers, discounts, and write-offs on all medical billing — for my cancer treatments, for Chris' arthritis treatments, and for other procedures. For example, before they would perform carpal tunnel surgery on Chris' hand, we were required to sign a form that stated that the whole procedure cost about $5000 **and we recognized that we were responsible for that full amount.** When write-offs and discounts were factored in and all was totally completed, our share was less than 10% of that number. Had we no insurance at all, however, or had we only poor insurance and a high deductible, we would have been responsible for the full $5000.

There appear to be contingency programs available to help patients with no insurance meet their payment obligations. We didn't learn details of those programs because we have health insurance, but we saw notices that announced the presence of such programs. Without any health coverage, one must decide whether each procedure is absolutely necessary and if one wants to spend the required amount of money to obtain it.

For example, if we had no insurance, we would have had to decide whether Chris **really needed** the a $5000 carpal tunnel procedure, or whether I **really needed** $500,000+ chemotherapy. Fortunately, we didn't need to make such decisions.

In our case, all *routine* costs are covered by Chris' primary health plan. Further costs, co-pays, and co-insurance have been covered by our finances aided by our supplemental plan.

The Lord Knows!

I would like to suggest that Chris knew her food service job would come with good health insurance and I knew what I was doing when I purchased our supplemental policies — but I can't. We didn't know!

We really **cannot** take credit for smart decisions. I'm sure if Chris did anything right when she took the food service job, and if I did anything right when I signed up for the cancer policy, it was because the Lord was guiding us. He knows both of us and He knows that we want to do His will. Occasionally, He may need to jolt us to get our attention, but He knows ultimately that our will is to serve, follow, and obey Him.

Any credit for smart moves and good planning must go to the Lord. He knew, even when I didn't yet have cancer, that I would be dealing with it in the near future. He knew, even then, what our needs would be. He was greasing the skids, even then, and starting to put everything in order to prepare us for this season of cancer diagnoses and treatments.

He certainly knew what our financial state would be, and He was preparing us for that, too. I remember a day back in July (after diagnosis and after learning that I would be disabled) when I was worrying about our diminishing finances. A letter in that day's mail and a phone call jogged me back into reality. They reminded me that the Lord is in control, and that I am more important to Him than the lilies of the field, which He keeps in fine array! (Matt. 6:28-30)

Our transition took us from **talking about** living by faith (before diagnosis) to **actually** living by faith (after diagnosis) over the course of a single day. I'm sure the Lord had spent years preparing us, but when it actually happened, we certainly weren't ready for it. I had no choice about getting cancer and I still have no choice. I can grit my teeth, concentrate as much as possible, set my mind to it, etc., and it won't add one iota of coin to my pocketbook nor will it make my body cooperate the slightest bit better with the cancer drugs, nor will it change any diagnosis. The same holds true for my wife and her medical problems. All of this is totally out of our control!

But we don't need to worry — the God of the Universe, who loves us, is in charge. He has been taking care of everything. When necessary, He has provided us with financial support, sent us meals, provided me with drivers, provided moral support and encouragements, provided a really broad network of Christians who are praying for us, etc. He has seen to it that Chris has a job with good health insurance, and that I have good supplemental insurance — we are covered! It is a chore just to remember all the things He has done for us — yet they all happened exactly when needed!

I find it difficult to imagine that anyone would be praying for me or my family, yet I know that through God's grace, numerous people (friends, family, students, acquaintances) are praying daily for us. They have all been holding us up in their prayers. It is incredible! Actually, it is quite believable (credible) because God is in control!

5

My Cancer – Multiple Myeloma

On 3 July 2008, I was diagnosed with **multiple myeloma**. According to the Multiple Myeloma Research Foundation (www.multiplemyeloma.org), "Multiple myeloma (also known as myeloma or plasma cell myeloma) is a progressive hematologic (blood) disease. It is a cancer of the plasma cell, an important part of the immune system that produces immunoglobulins (antibodies) to help fight infection and disease. Multiple myeloma is characterized by excessive numbers of abnormal plasma cells in the bone marrow and overproduction of intact monoclonal immunoglobulin (IgG, IgA, IgD, or IgE) or Bence-Jones protein (free monoclonal κ and λ light chains)."

Some cancers can be surgically removed. After surgical removal, chemotherapy, and/or X-ray treatments, the cancer can go into remission. Sometimes, however, the cancer returns, metastasizes, and spreads throughout the body to land elsewhere. Many have died because their original cancer returned, spread throughout their bodies, took residence in vital organs, and ruined those organs.

In my case, multiple myeloma is a blood borne cancer that resides in the blood stream, bone marrow, and bones **throughout the whole body**. It cannot spread any further because it is already **everywhere**. No *localized* treatments are available for this type of cancer.

Radiation treatments are of no general use for multiple myeloma because they would have to irradiate the whole body to reach everything — and that's not an option. Radiation can easily be used on specific locations (like the hair line fracture of my clavicle.) Ten 34 second treatments focused narrowly on my clavicular fracture completed that series. The radiation killed the cancer at the fracture site, and when that happened, the pain went away, and the bones were able to heal.

The treatment of choice for multiple myeloma is chemotherapy, which can reach everywhere in the body, while still being targetable in the sense that it can perform very specific tasks on very specific types of cells. Chemotherapy can adversely affect unwanted cancer cells although, **to a lesser extent**, it also adversely affects normal cells. This highlights the real trick for today's cancer research: identify medications that are detrimental to cancer cells while allowing all other cells to continue to function in uninterrupted normal fashion.

Average Life Expectancy After Diagnosis

The Multiple Myeloma (MM) literature describes it as an *incurable* disease that begins from *unknown* origins. Fifty years ago, the average life expectancy for MM after diagnosis was less than two years. When we asked our oncologist how long one could live with MM without any treatments at all, he basically suggested that a person in Stage 3 (of 3, which is exactly where I was when diagnosed and when we asked the question) would live about a month without treatments — and that would be it.

Our family doctor told my wife that MM produces painful deaths because the bones of the body become weak and break easily. I know from experience that such breaks are painful, so to have breaks begin to occur all over one's body for little apparent reason would be terrible. The hairline fracture in my clavicle was awfully painful until the X-ray treatments relieved the pain. I can only imagine how multiple fractures scattered around one's body would feel. A person's movements would necessarily be severely curtailed, and the person would need lots of pain killers to achieve any semblance of comfort and normalcy.

The original medication they found to treat MM about fifty years ago was one that killed both cancer cells and normal cells. Before that, they would have been searching for any type of medicine that would help someone with MM (if they had any idea at all what was wrong with the person.)

Even when they first identified MM as a disease and found a medicine to attack it, to treat someone then would be to administer some medicine to kill some of the cancer cells (and kill some good cells too) to knock back the cancer's progress. Then, they would wait a little while, and do it again to kill off a few more cancer cells (and a few more good cells too), and wait again. Etc.

The research in the battle against MM has come a long way since those days. The most remarkable gains have come in the 1990s and early 2000s (i.e., within the last twenty years.)

Each person's case, obviously, is different. A Philippino friend of one of my former students has MM. After six months of chemo, her cancer is in remission. She was in bad shape in the hospital before they were able to diagnose the MM. We recently also learned that the father of a friend of ours died of MM in the 1990s. He was so far advanced when diagnosed that he wasn't stable enough to even begin chemotherapy.

When diagnosed, I was in Stage 3 (of 3.) I was also told that I was a "healthy, young male" of 60, and as such, I was a candidate for the most aggressive treatment available. Sixty is young, relatively speaking, when the points of comparison are frail 75^+ year old people. Throughout treatment, everyone at the doctors' offices and clinics constantly told me I looked great and was doing fantastic! They were comparing me to others who routinely come through their doors — so, relatively speaking, I looked pretty good.

Back to the average life expectancy issue: The day I was diagnosed with MM, they handed me literature about MM and cancer treatments. It was the usual stack of stuff one would expect to introduce a new cancer to a new patient. The page that spoke of average life expectancy said that the current average life expectancy after diagnosis was 4 years. Forty years of research had advanced the average life expectancy by about two years (which isn't very promising.) At the time we read the *four years* number, we weren't yet familiar with the value of *two years* from years earlier. My oncologist seemed very optimistic that first day, although his optimism didn't totally square with the number 4 on the page. That same day, he told us, "I know what cancer you are dealing with and I know how to treat it." Now **that** was reassuring!

A week later, when we had calmed down a little bit, we met with him again. This time, we asked about average life expectancy after diagnosis and he said the current values were "ten plus years and growing" for the most aggressive treatment method. And as a healthy, young individual of 60, I was a candidate for that treatment.

I noted that the oncologist was obviously in a hurry to begin my treatments. Had he been able to schedule it, he would have started my first chemotherapy treatments the following Monday, which was only eleven days after diagnosis. As it turned out, he could not get all the preliminaries scheduled properly that first weekend, so my first intervenous (IV) chemo treatments started 18 days after diagnosis. He submitted the prescription for the main oral cancer drug on the day of diagnosis, so I was able to begin that regimen within 12 days of diagnosis.

The average life expectancy had gone from two years (fifty years ago) to four years (only 4 or 5 years ago) to greater than ten years in 2008, and the greatest improvement in average life expectancy after diagnosis only occurred within the last few years.

More recently we learned from the bone marrow transplant doctor that his colleague has been using this **most aggressive** treatment for more than 15 years, and some of his early patients have been in remission ever since their treatments. As they keep monitoring those early patients, and their cancers stay in remission, the average life expectancy for that treatment group continues to rise. That is good news! Actually, that's not only good news, that's GREAT NEWS!

Significant advances have been made in MM treatment regimens with the greatest gains occurring most recently. As an engineer, scientist, and computer geek, it is absolutely amazing to see the great numbers of advanced computerized instruments to which I (or parts of me) were attached, subjected, probed, irradiated, and/or analyzed since diagnosis. I know from personal experience that most such devices were not available 20 years ago. Twenty to thirty years ago, I was encouraging the ceramics industry to use more computers in their labs and plants, and I was told by friends that I was ahead of my time. That was in the early 1980s!

Analytical instrumentation companies have pushed the state-of-the-art a lot since those days because now all analytical instruments are fully computerized. Twenty years ago, if you wanted a fully computerized analytical instrument, you had to automate it yourself. I could do it. Few others were interested. When fully computerized instruments finally became available on the market, older non-computerized dinosaurs began to disappear. So I know from experience that most of the instruments and devices I have seen in hospitals, testing labs, and operating rooms during my treatments are new and generally represent today's 'state-of-the-art.' I'm only sorry that I was not awake during some of those procedures so I was not able to watch as they used the devices. I paid close attention when I was awake and was allowed to see what was happening.

We Americans of the 21st Century are so blessed to be beneficiaries of the computerization, advanced instrumentation, and incredible capabilities of medical science available today. So blessed! And advances such as these have facilitated the recent rapid advances of treatments. We can now do more procedures more precisely than ever before; we can now do procedures we were never able to do before; and we can now analyze and learn what is happening at molecular levels that we struggled to identify, label, and understand before. Research results are multiplying rapidly as a result of the advanced technologies available in today's society.

For example, about the time I was starting chemotherapy, a former student of mine was undergoing minimally invasive open heart surgery using a robotic system. They fixed her leaky heart valve and two weeks later, she was back at school, active, and functioning normally again as if nothing had happened. She has no zipper scar up her chest or anything like that. They inserted the robot arms through three holes in her right side to fix her left heart valve. It was not quite an outpatient procedure, but in comparison to normal open heart surgery, it was very close to it. She was in the hospital for days — not weeks or months. This is a relatively new technique — thanks to recent technological advances!

How fortunate we are to live today with all of the wondrous capabilities for analysis and treatment available to us! **Cures** for MM and other cancers should soon be forthcoming!

What is MM?

Multiple myeloma, in layman's terms, is a cancer of the blood plasma. It is a mutation of a particular type of plasma cell. Once this type of cancerous cell forms, it begins to multiply, and multiply, and multiply, and multiply, etc. Its numbers grow and grow and grow and grow.

What does this particular type of cancerous cell do? Nothing!
It has no apparent good use at all! But once it is present, it multiplies on
and on and on.   And it appears that its reproduction process has no
obvious OFF switch.   Once it begins to reproduce, it is very difficult
(read: almost impossible) to stop.

The whole problem appears to be a simple numbers game.
Unchecked, cancerous MM cells can grow to such great numbers that
they crowd out good cells and cause major problems by their sheer
numbers. As their numbers increase and they need to find more places
to live, hide, and multiply, they make space for themselves in the bones
of the body.

These cancerous cells spread through one's blood stream and into
the bones where they then cause other problems. Bones are porous to
start with, but to make even more space, the cancerous cells kick calcium
out of the bones to form lesions (which are essentially large holes – visible
in X-ray scans) in which they can sit, hide, and continue to multiply.
Each such lesion becomes not only a storage space for the cancerous cells,
but a weak spot in the bone. As lesions scatter throughout all bones of
a person's body, the person can begin to have 'multiple' myeloma
problems. Hence the name *multiple* myeloma.

In extreme cases, one can have so many of these cancer cells in
one's blood stream that the blood viscosity thickens. This is an indication
of a crowded system. And as blood thickens, the heart must work harder
to pump and circulate the required numbers of good blood cells
throughout the body. When blood is overcrowded with huge numbers of
"do-nothing" cancer cells, there aren't enough normal cells circulating for
the body to function properly.

When the heart must pump harder and faster to maintain proper
flow of thickened blood which contains lower percentages of good blood
cells, everything begins to deteriorate. The liver and kidneys then, too,
become susceptible to malfunction and blockage because of the sheer
numbers of cancerous cells in the blood stream.   The liver can
malfunction and kidneys can become physically blocked as they try to
filter and remove unwanted, unneeded particles from the blood stream.
And on and on it goes ....

The only thing these cancer cells seem to be able to do is multiply. They do so *ad infinitum* and that means they continue to multiply until they overwhelm every other part of the body.

These cancer cells multiply the way a gambler wants his chips to multiply. Play them and let the winnings ride. Each win doubles their numbers. A few instances of successful doubling can very quickly produce huge numbers of chips. These cancer cells multiply like that, too. They multiply quickly and they spread throughout the body as they continue to multiply — even though they are functionally useless. Their enormous numbers cause great problems in the body by overcrowding and gumming up the works of the body's normal functions.

The goals of today's MM research are to learn how to disrupt the automatic multiplication of these cells, to identify their reproductive OFF switch, and to help the body's normal functions to kill and remove these cancer cells from all body cells. None of these is particularly easy to do, and that is the specific nature of the problem! Recent advances in technology, however, have helped to produce wonderful new research successes!

Section 2

Chronology

6

The Preceding Year &
Discovery of the Cancer

If I think back to when and where I can first identify any symptoms of this cancer, I need to go back almost a year before my date of diagnosis. I'll start there.

A Sneeze

Back in August of 2007, Chris and I were dancing in an exhibition in Greenville, SC. We were dressed and ready early, so we stopped along the way at a local shopping center. We went in to browse through a fabric store.

In the store, I sneezed. It was a particularly powerful sneeze accompanied by an equally powerful burst of pain in my left rib cage. It felt like I had been punched in the chest and had broken a rib — and all I had done was sneeze. I immediately had to steady myself and then sit down. I returned to the car, reclined the seat, and tried to relax until Chris was finished shopping. The next question was whether or not we could continue with our dance exhibition. Certain movements brought back the pain at that spot, but we went to the exhibition, the dance went on, and we completed our performance without incident.

Following that event, I was very careful with movements that might exacerbate the problem (whatever it was), and I was especially careful whenever I again felt I had to sneeze. Sitting quietly was most comfortable, so trying to relax and not move that part of my body became my primary mode of treatment.

Over a period of a few weeks, the pain subsided and disappeared completely. End of story ... or so I thought.

Several Months Later

At Christmas time, the pain started to return. I assumed I had done something to re-injure the rib and re-aggravate the problem, although I had no distinct memories of any such incident. The pain had returned, for no apparent reason, and I began to deal with it again – similarly to before. I tried to not move my rib cage any more than necessary. That meant I sat around a lot, moved slowly and carefully, and was generally very cautious and observant regarding my left ribs and whole rib cage.

Self-Treatment:

Minimize Movements

All the while, I was paying great attention to the way I felt and to the locations of the pains. I had concluded with a fair amount of certainty that I had actually cracked a rib. That explained both the pains and the location, and it answered all questions quite well. So my self-diagnosis was *cracked rib*. My self-treatment was to *minimize movements* (i.e., move slowly, sit quietly as much as possible, and don't do anything to aggravate the problem.) If and when necessary, I would also take over-the-counter pain killers and use a heating pad.

Both my wife and the algebra teacher at school encouraged me to go see the doctor. I told them both that he would charge me $50 to tell me to go home, to take it easy, to not aggravate the problem, and to take pain killers as needed. Since I was already doing all of that, I waited.

As the Spring Semester passed, it became more and more difficult to move. I found a comfortable recliner at school in the library and I sat in it each morning during devotions. Getting out of the chair, once I was in it, became more and more difficult because I had to bend my body forward and slide forward before I could stand up. All of that, which sounds easy, and normally is easy, became more and more difficult. The pains worsened progressively throughout the semester.

Try Not to Sneeze or Cough

It was also becoming much more painful every time I coughed or sneezed. I especially dreaded sneezing and fortunately, the Lord allowed me to have a long period of days in which I did not sneeze or cough at all. When I did sneeze or cough, it felt like my whole rib cage was ready to explode. Actually, it felt like there was no flexibility in my rib cage. It was like blowing up a balloon, but the balloon was already at its size limit. The extra pressure didn't change the volume of my lungs or rib cage, but it produced pain by stressing the rib cage. I learned to hold my arms tight to my chest when I felt a sneeze or cough coming to try to minimize the pain. That worked for a little while, but didn't solve the problem. Things continued to get worse.

Pay Attention to Locations and Causes of Pain

I had been paying close attention to the locations of the pains from that rib. I followed the pains to the front of the body when my sternum started to hurt. The injured rib is attached to the sternum, so that made sense. Then, I had similar pains down around my side and towards my spine from that same location. That made sense too. Everything from sternum to spine along that rib was attached — so the pain locations seemed logical.

But then, I began to feel pains lower down in my abdominal muscles and lower down my back. Those locations were not directly attached or related to the location of the injured rib. So those new pains threw question marks into my self-diagnosis. Then, strangely enough, I began to feel similar pains at the mirror image locations on the opposite side of my body.

For example, one day I bent over funny to the right and thought I had stretched my left oblique muscles. Over the next day or so, I began to feel soreness in my lower left side and I was careful to not aggravate those muscles any further. I used medicated patches or the hot pad to try to help those areas. After a few days, the pain would subside and disappear. But then, for no apparent reason, I would begin to feel similar pains on my right side near the same location. I could usually identify one event that might have caused the pain at one spot on one side, but I was not able to identify why the mirror image location was also having similar problems. That made no sense at all.

Locate Comfortable Chairs, Sit More, Stand Less

All this while, the discomfort was slowly intensifying, moving became more difficult, and getting in and out of chairs was more difficult as well. I had located all possible comfortable chairs in which to sit to minimize the discomfort. To sit quietly produced no pain or discomfort, but eventually, I always had to pull myself out of the chair and that always reminded me that I had a problem.

Standing and sitting were relatively pain- and discomfort-free. Changing positions between the two was when I could feel the pain. Standing for long periods of time in the class room became more and more difficult. Fatigue caused me to slouch more, which stressed those muscles and bones and brought more discomfort. As it became more and more difficult to stand at the front of the classroom for whole class periods, I began to sit more. Fortunately, we all had tall stools with arms near our lecterns, so it was possible to sit in the stool at the lectern to teach. Actually, I had two stools by my lectern. I sat in one and propped my feet on the legs of the other, so I was able to continue to find comfortable positions at the front of the classroom. As the semester moved along, I spent more time sitting, less time standing and walking, and I moved even more slowly to not aggravate the muscles in my left side.

Eventually, I had pains around my rib cage and abdomen for seemingly **random** reasons, so the *don't aggravate* idea wasn't working. I wasn't aggravating anything and the discomfort and pains continued to worsen.

Go to the Doctor

During April of 2008, I began to search my medical CD and the internet to determine what could be causing the problems. Finally, I scheduled an appointment with the doctor. The night before the appointment, I came across *costochondritis* on the CD. This appeared to define my problem. Costochondritis refers to an inflammation (the *-itis* part of the word) of the costochondral region which is the chest/rib cage area. I had never heard the word before, thought it was strange, but wrote it down and stuck the note in my wallet.

The next day when I went to the doctor's office, the nurse interviewed me first, and after playing twenty-questions, I told her that I had written down the name of the ailment from my medical CD that best fit my symptoms. I pulled out my wallet and showed it to her. Then, she flipped over her clipboard to show me the same word as her suggested diagnosis.

When the doctor came in, we talked. At the end of the discussion, he too concluded that costochondritis was the most likely diagnosis. That turned into a $50 office visit with instructions to continue to minimize my movements, to not aggravate the problem, and to take a specific over-the-counter pain medication. My statement from months early had come true. I had waited more than three months for the doctor to finally tell me that he agreed with both my self-diagnosis and my self-treatment. He had poked and prodded me around the chest, back, and abdomen, listened to my heart and lungs, etc. – you know – all that good doctor stuff – so he had earned his $50 (or however much it actually cost me), but he hadn't provided me with much of anything that was new or helpful to solve the problem.

When I left the office, I was happy that he had agreed with my diagnosis, yet sorry that he couldn't pinpoint a better, more specific problem.

The Injury to My Shoulder

DSL Line Problems

A few weeks later, our DSL internet line was acting up and the boys were complaining. The phone company was responsible for the line to the outside of our house and I was responsible for everything inside the house, so every time we called the phone company to report a DSL problem, they always sent someone to check our house's **inside** wiring first. Our wiring had been checked three of four times already by three or four different technicians — all of whom made the same checks, all of whom ran the same diagnostics, all of whom proclaimed our inside wiring to be OK, and all of whom said our problems were **outside** in the lines leading to our house. The problem had something to do with the local squirrels liking to eat cable insulation which then allowed moisture into the cables which disrupted communication signals.

That day in May was no different. The serviceman came in to check our **inside** wiring and modem and he repeated all of the same inside checks. Then, he asked to see where the inside wires connected to the outside wires. That location was above the drop ceiling in our downstairs bathroom. I pointed to that location and he removed a ceiling tile to expose the wires. It was dark up there. I couldn't see the wires easily and I figured he couldn't either, so when he went outside to get something from his truck, I climbed up on the chair we were using as a ladder to remove the next tile and fully expose the connections.

The Fall

While standing on the chair, its one leg buckled, and I came tumbling down. In that instant I grabbed for anything I could reach to stop my fall. My hand grasped one of the aluminum support beams that holds the drop ceiling. As everyone knows, those aluminum beams are not particularly sturdy. They are designed to hold very light-weight stationary tiles. But here I was, a large man falling off a chair, with my right hand grasping the exposed aluminum beam.

It couldn't stop my fall completely, and it didn't — but it slowed my fall. Instead of crashing into the wall and toilet with my whole body, I was able to stop myself against the wall and toilet without actually crashing into anything. The aluminum beam was twisted royally out of shape, but I had not broken anything. The chair's leg was bent, but the toilet, wall, and my body were intact. Had I not been able to grab the aluminum beam, I fear that the toilet, the wall, and my body would all have been broken into pieces. I'd have hit the toilet and the wall with the exact part of my body – my left side – that I had spent the whole semester trying to protect. But I was saved by the light weight aluminum beam! (Believe it or not — it's true!)

I quickly regained my feet, twisted the chair leg back into proper position, twisted the aluminum beam back towards some semblance of straightness, and removed the ceiling tile so the serviceman could see the wiring easily.   When he returned, he examined everything and proclaimed it all **OK**. The problem was **outside** the house, as we already knew. So he reported the problem (again), checked outside for a while, and left.

The reason for this story is that over the next several days, I began to have pains in my right shoulder joint. I hadn't done anything to my right shoulder that I could remember, other than that I had yanked it well when I fell off the chair. I hadn't jarred it or slammed it against the wall or against the toilet bowl or anything like that. I had simply yanked it (stretched it hard and fast) when I fell, clinging to that little aluminum beam. That was the only event I could possibly relate to the pain in my shoulder.

A Return to the Family Doctor

I tolerated this pain for a while, and finally, without any relief, I set up another doctor visit for the shoulder pain. This exam was scheduled a few days after my last class of the school year in early June 2008.   At this appointment, the doctor examined the movement capabilities of my arm and shoulder joint, determined the limits and angles I could achieve, tried to identify positions which produced pain, etc., but he eventually couldn't tell me anything further about my shoulder other than that I had somehow stressed it to cause the pain.

Then he asked about my chest problem. At the earlier meeting, he had told me that if the problem continued, come see him again and he would look into it further. I had procrastinated beyond the two or three weeks he expected it would take the over-the-counter pain reliever to solve the problem, but by this time in June, the pains were worse. So I told him that the problem was worse and I requested a stronger pain killer and a better anti-inflammatory drug. He gave me prescriptions that day for both.

But he didn't stop there. He began to perform blood tests over the next several days — all sorts of blood tests. Every few days, I would get a call from one of his nurses requesting my presence so they could draw more blood samples.

The Diagnosis and Referral to An Oncologist

My next appointment was scheduled **after** the 4th of July, but since my pains were still getting worse, I rescheduled for 3 July 2008 — the day before the holiday. Finally, maybe, I would find out what was going on.

When I returned that day, the doctor spent a lot of time going over the blood results showing me which parts of the blood analysis pointed to the proper functioning of my various organs.

Then he pointed to an **M-spike** number on a special blood test. He indicated it was high and it might be pointing to a cancer problem. Anything else he might have concluded from the blood work became secondary to dealing with that most important finding first: a possible cancer. So he said he was going to refer me to an oncologist. He was sending me to the oncologist he would go to if he had the same problem, so I was going to be in good hands. Although the diagnosis was a shocker, his comments about the oncologist were very comforting.

He left the examination room, went across the hall to his office, and called the oncologist. He left both hallway doors wide open, so I could hear his half of the conversation. I heard him repeat numbers off of the blood analysis sheets, and I specifically remember him giving the **M-spike** value. When he returned to the examination room, he told me the oncologist wanted to see me right away at his office in Greenville — come right up — he would work me into his schedule. So the family doctor wished me luck, and I was off.

All of this happened first thing in the morning of the day before the holiday. Chris was off work and still sleeping. I called immediately to wake her up with the news that we had to drive to Greenville to see an oncologist because the family doctor thought I might have cancer.

I admit — that's not a great way to wake up your wife, but I wanted her to go with me, and I wanted to get there ASAP to find out what was actually wrong with me.

97% Certainty

When we met the oncologist, he said that he had talked to our family doctor and he had already seen the faxed copies of the blood work.

From the value of the **M-spike** number alone, he said, "I am **97% certain** that you have a cancer called ***Multiple Myeloma***." He then scheduled tests during the following days to confirm his diagnosis.

As an engineer, I knew that if I had claimed 97% certainty about something, there would be little doubt in my mind. The 3% was for the tests to confirm the diagnosis. The 97%, for all practical purposes, meant he was **certain** I had cancer.

That was how I learned about my multiple myeloma (MM.) It was a long tortuous nine-month period of cracked ribs, discomfort, pain, minimized movements, etc. — all of which were accompanied by increasing discomfort and pain.

During that first visit to the oncologist, he started the process needed to prescribe the main oral cancer drug for MM, which was *thalidomide*. That same day, he also prescribed even stronger anti-inflammatory and pain killer drugs. Since the 4th of July weekend started the next day, he said he wanted to make sure I had some comfort over the holiday. He gave me the prescriptions, we filled them, I took the pills, and they worked.

The next persons to ask me how I was doing received the answer, "I have cancer." Without question, the identification of cancer was a major shock to both Chris' and my systems. But we were spending the 4th of July weekend with two of her brothers and their families, so we could talk about it and begin to absorb the diagnosis.

That weekend, we spent lots of time discussing cancer and cancer-related issues. We had lots of support from everyone throughout that weekend.

My Christian Outlook on this Problem

As a Christian, this chronology wouldn't be complete without mentioning another point-of-view that overlapped the events I described above:

10 Years Left

According to the Bible, the normal length of life is "**threescore and ten**." (Psa 90.10) So when I awoke on my 60th birthday on 25 June

2008, I announced to Chris that I had 10 years left. Seventy minus sixty is ten.

I hadn't really ever spent a lot of time considering the length of my life or how many years I had remaining. It is all in the Lord's hands and I simply hadn't thought much about it. I still haven't thought much about it, but when one gets to nice round numbers like 60 and 70, the arithmetic is simple and it is easy to say, "OK – ten left!"

I was actually trying to be somewhat funny by making the statement, although I was also realizing that approximately 6/7 of my life was behind me and 1/7 remained. Depending upon how much you want to dwell on stuff like that — such thoughts can be very sobering.

As a Christian, without any major health problems, I hadn't ever given it much thought. My concern had been that the Lord was using me in the now. The future would take care of itself.

So that was on my 60th birthday, 25 June 2008.

4 Years Left

A week later on 3 July 2008, I received the diagnosis of cancer. On that day when we first met the oncologist, he gave us stacks of papers and information describing multiple myeloma. One of those sheets said that the average life expectancy after diagnosis with MM is four years.

That number would require a change to the statement I had made on my birthday. But rather than think about it too much or get upset, I ignored it.

I know that the Lord is in control of my life and the lives of all of our family and friends, so it is His decision when our times are up. Average numbers like these are guidelines — averages — but not numbers cast in concrete. You or I could die today and such events would alter only slightly the new averages. The point is: such numbers are guides. The controlling factor is the Lord. I am the Lord's. I know that He will decide when it is time for me to depart this life. It will happen when He says so — no sooner — no later. So it is not an issue with which I have ever been overly concerned. I don't have any control over it anyway, so worrying about it accomplishes nothing.

10⁺ Years & Growing

Another week later, on 11 July 2008, we again met the oncologist in a somewhat more relaxed meeting, and he said that since I am considered to be a "young and healthy" individual, I was a candidate for the most aggressive treatment for this cancer. Average life expectancy after diagnosis with the aggressive treatment is "10⁺ years and growing."

So over the course of two weeks, my life expectancy went from 10 years according to the Bible, to 4 years according to a 3-4 year old information sheet on MM, to 10⁺ years according to latest results relayed to us by the oncologist. I was back to where I started on my 60th birthday. Nothing had changed.

The Lord Is In Charge

I know that the Lord is in charge of my life, so I have never really dwelled on how long I will live, but on whether or not I am fulfilling the Lord's will for my life. I am probably not doing a very good job of that, but **it is my desire to fulfill His will.** I want to be doing what He wants me to be doing. I want to step through doors He has opened for me. It is not always perfectly clear where He is leading or why, but it remains for us to be patient and to follow His lead. That has been my goal. I fall short like everyone else, but at least I have set that as my goal — and I believe it is a correct goal for each of us to have in our lives.

Who knows in more detail the precise natures of our talents and gifts than the God who provided them? Who knows most precisely, therefore, the best possible places for us to put those talents and gifts to useful purpose than the God who provided them? Who knows more about our moment by moment needs, and the moment by moment needs of others, than our loving God?

We put our lives into such incompetent, fragile hands when we attempt to control everything ourselves, separate from God's guidance. The Bible says we aren't capable of guiding our own lives (Jer. 10:23) and from personal experience, I know that is true. We simply don't have access to all the pertinent facts to make perfect decisions. But God does. He knows in great detail what is happening in each of our lives at all times. So we should look to Him for all of our needs.

I have chosen to trust Him for this. He has never failed me. Looking back, I can recognize places where I would have made major blunders in my life or in the lives of my family if I had made certain decisions, but God was in charge. I can also recognize places where I made decisions that at the time appeared to be consistent with God's will — that now I am not sure were the right decisions. It's too late to correct or change those decisions, but I'm patiently waiting to see if God will show me at some point, "Yes, that was a good decision," or "No, that was a major blunder." Patience is not always easy, but I'm still learning.

God is good. My life and the lives of my family and friends — all our lives — are in His hands. We are imperfect and we blunder regularly, but to the extent we allow God to control our lives — He can and will take care of us.

During this time of cancer and medical treatments, I gave God control of everything important so I could turn my focus inward — on a minute by minute, narrow basis. I didn't have the time, understanding, or power to control the big things. He did. All I could do was pay attention to how my body was reacting to specific treatments — and accurately relay that information to the medical staff. Every time I moved, I could feel my body reacting in ways I had never experienced before. I paid close attention to what was happening to me, what the chemo treatments were doing, and how I was feeling minute by minute, day by day. I was not dwelling on next year or five years from now. I was here and now.

Yes, I'm concerned about my wife and family. I don't want to be overly burdensome on them, but I know that during this time, I **am** a burden, whether I want to be or not. I know my cancer causes them stress — especially my wife. I cannot do anything about it. The cancer is here. I must deal with it. She must deal with it — and me. But we both know that we are in the Lord's hands, and He is in control. If we get to next year or five years from now and we are still functioning, He will still be in control and hopefully we will be doing exactly what He planned for us.

In the meanwhile, we read, study, learn, pray, and hope to grow in the knowledge of His Word, His will, His graces, and His mercies. There is a lot to be learned in times like these because we have nowhere else to turn and no one else to turn to.

Maybe this cancer was God's way to get my attention. I think He already **had** my attention and He was **preparing** me and us for this ordeal. I'm not totally sure — I'm waiting to learn more complete details from Him — but **I am** paying close attention.

The point is that we don't know when our lives will end. God does, but He's not telling. We are in His hands and under His control so it is not something that we need be overly concerned about.

You could argue that if I had been 35 when diagnosed with MM, whether my life expectancy was 10, 4, or 15 years, none of those numbers would have sounded good. But as a Christian at age 35, even then, I knew my life was in God's hands. When I accepted Jesus' gift of salvation a long time ago, I placed myself in His care — knowing full well that He was going to do a much better job of caring for me than I ever would or could do. I haven't been dissatisfied with anything He has done. Sure, I have hindered His ways by doing my own thing and throwing my will, desires, and decisions into the mix, but I know He is capable of handling me and moving me forward to the place where He wants me to be.

The best thing I have found along the way is my constant desire to do His will and to be functioning within His will. Without that desire, He would need to stop me much more often to get my attention.

I am at that stage in life, where I want to do His will all the time. It is a constant goal. I'm sure I fail miserably, but it is a constant goal. And I have learned just enough patience along the way to know I won't immediately get the answers to all of my questions. But with patience and attentiveness, I may actually learn something and be more useful to Him in His service.

When we have the Lord as our Master, it provides us an outside focus that can help us to look beyond our immediate problems — not necessarily to a time six months or six years in the future — but to a time when we are serving the Lord to our fullest extent. We can be working in His service from a hospital bed, or from a wheel chair, or from any number of other positions that natural man would label as positions of non-service. We can be working in His service as cancer patients, or as cancer survivors, or as cancer support group members, or in any number of other positions that society would label as unneeded, unimportant, or useless.

Society doesn't know nor care about the needs of individuals. God cares about each of us individually, and if and when we need help as individuals, He provides for those needs. One lady who answers the phone at a mail-order pharmacy spent a few minutes one day chatting with me about my cancer. She promised to remember me in her prayers that evening. That wasn't her job — she was supposed to arrange shipment and payment for my medications. She did that. But she also said something personal that encouraged me as an individual. Society would be upset with her for saying anything off-subject like that — but she might be working for that company **and** fulfilling God's will as she works for Him, too.

The point is that God knows fine, precise, detailed information about each of our lives and He has fine, precise, detailed control over each of us that mankind as a whole (and men and women as individuals) simply cannot and do not have. He knows each of our thoughts and needs and He provides for them. We don't clearly know our own thoughts or our own needs, let alone the precise thoughts and needs of those around us.

So who has the best capability for guiding our lives? Do we? No! God does!

I didn't know I had cancer. It was 10 months between when symptoms started to show and the actual diagnosis arrived. I didn't know. The doctor didn't know. But God knew. And when I think back upon those days, I can see that He was preparing me and us for this time.

This is further proof to me that we are in excellent hands when we are in the Lord's care! And that care is available to everyone, regardless of economic status or well being. He wants us all to come into His family and allow Him to be our loving Father. I accepted that offer and am a beloved member of His family.

He is my Lord. I trust Him completely. And I know He is always there for me when I need Him. Even as a consultant who spent months in Indonesia away from my family, who spent many hours alone after work on each of those days — He was there with me! I was never truly alone. And He is with me now! I trust He is guiding my fingers as I type. I trust He is paying better attention to how my body is recovering than I am. I trust that last fall, He helped my body react properly to the chemotherapy. He knows what He's doing and He has capabilities we simply don't have. After all, this is the creator of the universe we're talking about.

He is my Lord and He cares for me. It can't get any better than that. And He offers that same relationship and the accompanying watch-care to one and all. I hope and pray none of you have to go through ordeals like cancer and chemotherapy. If you do, I hope and pray none of you have to go through them without the Lord at your side.

Surely, it can be done without Him. Anyone can go it alone! But it is wonderfully comforting — even soothing — to know that the Lord is working for you — to support, strengthen, and see you through the ordeal — to know that He has a purpose for each of us and He will work to fulfill that purpose in our lives if we allow Him to do so.

7

Diagnosis – Disabled!

It was a big shock the day the doctors told me they thought I had cancer. Confirmation of that diagnosis took another two weeks, and the delay provided time for the impact of the diagnosis to sink in. My recollection of my reaction to the cancer diagnosis is one of stunned silence. It did not hit me like a 2" X 4" across the forehead, nor like being hit by a train, nor like running into a brick wall. It all seemed rather surreal. I don't remember any major response or outburst at all. Maybe I was so stunned, there was no overt reaction at all.

The week following the diagnosis, we visited the oncologist for our second visit. At that appointment, I asked how the cancer was going to affect my teaching. I remember the doctor saying, "Oh, you won't be teaching because you will be totally disabled!" **That revelation was much more of a slam to my being than the original cancer diagnosis!!**

I always understood that to be disabled as a teacher meant you had an enormous problem! To be disabled from a construction job or some other mostly physical job meant there were other, less strenuous, fall-back options for meaningful labor. But there is no easy fall-back option for a teacher. So there I was, listening to the doctor tell me that I had a huge problem!!

How temporary was the disability to be? He said it would be "permanent." (Not good!)

How was I going to respond to chemo treatments? He said he didn't expect me to get sick or nauseous from the treatments, but he expected me to be very weak — hardly able to do anything. I would sit around a lot. Looking back — he was 100% accurate.

During and after five cycles of chemotherapy, I was weak; my legs were sore; and I was not walking around any more than absolutely necessary. I now know what it means to be weak. Actually, I thought I knew what it meant to be weak before chemo started. Then, after adjusting my definition of *weak*, I thought I had learned what it meant after Cycle #2. But my energy after Cycles #3, #4, and #5 became progressively worse — until it was practically non-existent. It couldn't get much worse if I wanted to admit having any energy at all. I could still stand up, but it was a struggle. If it had gotten any worse, I would have been stuck in a chair all the time and that would have presented other problems.

As a teacher I needed to be able to stand in front of each fifty minute class, walk to and write on the board as necessary, and lead discussions. I could do most of that if I were sitting in a tall stool at the lectern, but to do so for 6 classes in a row each day would have been a major struggle. Looking back on it, I simply wouldn't have had the energy to teach.

My mind appears to be intact through five cycles of chemo, so that doesn't appear to be a problem. (My wife might disagree — so I haven't asked her opinion.) The mental parts of teaching don't appear to define the disability — the physical aspects do.

So I accepted the idea that I am disabled. I obtained a blue handicapped hang tag for the car which is very convenient because it minimizes walking distances from parking lots to stores. Many stores have really nice 4-wheel walkers with large baskets attached (i.e., grocery store shopping carts.) We have all used them for years, but they're especially great when one is weak and/or has walking problems. I can throw the cane in the basket and walk around the store while leaning on the cart.

As my energy level decreased, I requested a wheelchair. This allowed me to ride rather than walk. Even though I was able to walk throughout the whole treatment process, I did not have the energy to walk far enough to go to a mall or to walk through a large department store. I didn't want to go shopping if it required walking from chair to chair in the stores. A wheelchair solved that problem.

I quickly discovered that in addition to my legs not being strong enough to walk around a mall, my arms weren't strong enough to propel me through malls in the wheelchair. My wife has been kind enough to push me around and I continue to accompany her on her *shopping* trips.

The key word is *shopping*. She can spend a whole day trying on dresses and other clothes, trying on shoes and boots, and looking through all the racks of clothes and accessories without buying anything! That is how she defines *shopping*. And with the wheelchair, I can accompany her and not be stuck at home while she is off having fun.

So my teaching career stopped abruptly on that 11th day of July 2008 — and looks like it will NOT continue — consistent with the expression "total disability."

My Replacement at OCA

Let's back up a little bit. I am quite certain the Lord guided me to Oconee Christian Academy about two weeks before the start of the 2005-2006 school year. Within just a few days, I went from not knowing anything about OCA to being OCA's new science teacher.

Not really wanting to teach biology, I spent two years politicking for the 7th grade life science teacher to teach 9th grade biology. She seemed to like everything about biology and she would do an excellent job. After two years teaching biology at OCA, it was finally suggested that she might teach biology the following year. As I was about to leave school for the summer on the last day of the school year in June, 2007, the principal whispered in my ear that the new biology teacher might not be back in August. In other words, he was warning me I might be teaching biology again.

Sure enough, the life science teacher's husband accepted a pastorate in Greensboro, NC, so they moved during that summer. Once again OCA needed a science teacher — this time at the middle school level — so once again, they prayed about it and turned it over to the Lord.

A few days later, as the director was walking out the front door to his car, another fellow was walking towards the building. When the director asked if he could be of service, the fellow asked if they needed any teachers at OCA? "What kind of teacher?" "A science teacher." "What science specifically?" "Biology." Within a few days, OCA had a new biology teacher, and I was off the hook regarding the necessity to teach biology again.

Now we turn a year later to the summer of 2008 when I was diagnosed. On a Friday, I was told that I would be permanently disabled and I wouldn't be able to teach. I called the principal and requested he come over to the house Saturday so I could give him the news. That Saturday, in my living room, I told him that I was going to be totally disabled and I would not be able to teach during the coming school year. They were going to need a new high school science teacher.

Sunday, the principal told the director that I was no longer able to teach. Monday, the director called the other local Christian school to find out who they had teaching science. Turns out that the other Christian school had just merged with a third Christian school which allowed them to close the doors of their middle school/high school program. They weren't going to be needing their junior high/senior high science teacher, so they gave the director his name. He was invited to visit OCA on Tuesday. He was a retired engineer who could teach chemistry, physics, and calculus — which met their needs. Within a day or two, he was signed up as the new science teacher to replace me. The whole process of finding a replacement for me took less than a week.

The director later told me that he didn't understand how or why the new teacher had stayed in the area during the previous year. The other Christian school had only a handful of students who needed to study science. He had little if anything to do apart from those few students — but he nevertheless stayed in the area.

We know that the Lord had this all under control right from the beginning. I told the principal that the Lord knew I had cancer for quite some time. Therefore, the Lord also knew OCA was going to need a new science teacher, and He had made preparation for it. The issue was **not** going to be **how** to find a totally new science teacher for OCA, but to determine **who** the Lord had already selected and prepared for the job. All they had to do was identify the new person. Sure enough, within five days of my announcement, the new science teacher had been identified. OCA was all set!

All such stories, and these are only a few, gave great confidence to all of us that the Lord **is** in control — not only of our lives, but of the daily needs at OCA as well.

Joe's Schooling

Joe was attending OCA because tuition was free for children of teachers. What would happen when I was disabled, without an income, and no longer a teacher? Almost the first words out of the director's mouth after finding out that I had cancer was, "Don't worry about Joe. We'll take care of him. He will graduate with his class."

Not only had OCA stepped up to the plate to take care of Joe during his senior year, but the OCA family (staff, families, and friends) got together to help as well. Someone paid Joe's athletic fee so he could play soccer. The monies to cover a class ring, cap, and gown were picked up by another group of donors.

They took good care of Joe during his senior year. Several mothers as well as teachers sent word to us that they were "keeping an eye on Joe for us." One parent said that Joe was "one of my boys — don't worry — I'll take care of him." It is wonderfully comforting to see how the Lord works through His people.

I may be 60 years old, but I am still learning about certain aspects of Christian life that I never really experienced before. The learning curve has accelerated, though. At Clemson, they always talked about "life-long learning." Well, the Christian life is one of "life-long learning," and the learning continues regardless how old one may be. God knows what we need and sees that we are appropriately educated.

Living by Faith

Now that I am totally disabled and without a job, it is no longer a matter of waiting for the bi-weekly pay check to come in from school — while ignoring God's provision in our lives. During the first six months following diagnosis and during chemotherapy, I had no salary at all. We had to rely totally on the Lord's provision. Now we rely on a small monthly disability check and Chris' pay check. The biggest expense, the medical bills, are covered (for the most part) by Chris' health insurance from her job. We are now relying on God — not simply saying the words.

During the first six months when I received no income check nor Social Security disability check, we did not fall behind on any payments to anybody. The Lord provided for all of our needs. The checking account balance was close to zero once or twice, but we have always had money when we needed to pay expenses and bills. God is faithful!

One day in July of 2008, I was sitting around worrying about my coming financial status. How were we going to handle it? A few minutes later, the phone rang and later in the day, a letter arrived in the mail. The phone call regarded a possible source of income and the letter contained a check. God was telling me, "I know what's going on. Don't worry about it. I've got it under control."

Every so often, I received a large, nasty bill for my share of medical procedures. Timely checks arrived in the mail to help cover those bills. The Lord knows what is happening and He has taken care of our needs. Sometimes, the clinic's scheduler will call with a list of upcoming appointments, which include some expensive procedures. The doctors have been good about cancelling expensive, **un**necessary procedures. For example, twice they requested MRIs of my whole spinal column (billed at about $12,000 — my share of which is at least $1000) but twice, when we said we didn't have the money for such tests, they were cancelled. Expensive, necessary tests continue as planned and we trust the Lord will make sure we have the money when needed.

Living by faith and trusting the Lord for finances is still easier said than done. Now I receive a monthly social security check for my disability. It doesn't cover all expenses, but we know God has everything under control.

8

Diagnostics

Multiple myeloma is a blood plasma cancer. There are several main analyses methods that can be used to prove or disprove its presence. One is bone marrow testing. Others include blood and urine tests.

Many of the cancerous cells which are overabundant in multiple myeloma patients reside with the body's bone marrow inside bones. From that location, they can do lots of damage. Since they are part of the blood plasma, the cancerous cells can travel with the blood throughout the whole body. It is much more difficult to sample and analyze bone than bone marrow. So the bone marrow biopsy is a major test used to prove the presence or absence of the cancerous cells.

Other common tests, such as blood and urine analyses, are performed frequently to monitor blood counts, ions, proteins, and the presence and concentrations of cancerous cells. Samples for these tests are easy to collect and analyses are routine.

One blood electrophoresis test measures the **M-spike** content, which is an indicator of the number of unwanted cancerous cells in the blood plasma. That particular test is less commonly performed, but when a physician suspects cancer, it can be ordered.

Routine blood tests measure the numbers of white cells, red cells, platelets, hemoglobin content, etc. Chemotherapy targets cancer cells which reproduce quickly. But chemo also affects normal (good) blood cells. Routine blood tests monitor the body's daily responses to chemo treatments and the blood's recovery progress after each.

Some drugs cause stem cells to form quickly and enter the blood stream. The effectiveness of these drugs is also monitored with blood tests.

The nastiest test routinely used on MM patients, however, is the bone marrow aspiration. MRIs, ordered because they can show the extent of bone damage caused by the cancer, are expensive but **not** painful.

Bone Marrow Sampling

Since many cancerous cells in multiple myeloma patients reside in the body's bone marrow, and since bone marrow is **relatively** easy to collect and analyze, sampling and analysis of bone marrow is common in blood cancer patients.

The bone marrow sampling procedure is nasty, however, because it is *painful*. How painful is a function of the skill of the doctor taking the sample, the amounts and types of anesthetics used, and of course, the response of the patient to the pain killers and sampling procedures.

My first bone marrow sample was taken using only a local anesthetic. No other anesthetic or pain killer was offered and since it was my first bone marrow sampling, I had no idea what to expect or what I was about to experience after entering the room.

I now know that you can request extra pain medications before the process. I suggest that extra pain medications be requested prior to the test. If you need to have your bone marrow sampled, remember to discuss the procedure and available options with the staff ahead of time.

The Procedure

The typical procedure for bone marrow sampling is for the patient to lie on their stomach so the doctor or nurse practitioner has easy access to the upper part of the pelvis. They use a sampling spot on your back, a few inches below the waist, and a few inches to the side of the spine. In this position and at this spot, there is relatively little distance between the surface of the skin and the pelvic bone.

The area around the sampling site will be numbed with local anesthetic. The anesthetic is applied from the skin's surface down to the bone.

Then, using a large diameter "needle," an entry hole is made into the pelvis until the pelvis' central bone marrow cavity can be accessed.

At that point, using a hollow needle, they can suck the bone marrow (a liquid) into a syringe to collect the sample. This is the painful step.

Speaking from the experience of having undergone several bone marrow sampling procedures, the "needles" and sampling tools are very sturdy with 'T' handles at the top. Their "needle" resembles one of my shop tools. I am not sure of all details of this procedure because I've never been able to see what they were doing. I do know how it feels, though, and it isn't a fun test.

My First Sampling

During my first bone marrow sampling, the nurse practitioner who performed the procedure and her helper were discussing the proper needle to use. Picture this: I am lying on my stomach facing the wall while they are behind me where I cannot see what they are doing. Then I hear one of them quietly say to the other, "Do you think we need to use the lonnngggggg needle or will the short one be OK?" Neither one of them volunteered to show me the needles being discussed, so my imagination filled in the blanks.

To my mind, a "short" needle is about ½ inch long and a "lonnngggggg" needle is 6 or more inches long. And, of course, a "needle" is a needle, not a ¼" diameter hollow tube with a 'T' handle and cutting teeth on the circumference of its tip. I had all sorts of wild pictures running through my mind as they were talking behind me, and taking samples.

We need a side excursion to explain the images I was seeing: When I was in public school about 50 years ago, I was hospitalized to have my tonsils surgically removed. The boy in the bed next to me was having his appendix removed. During that hospital stay, I saw the size of the needles and syringes they used on me, which were all relatively small (i.e., normally sized). Even the large syringes didn't seem too large. But the day they were going to operate on my roommate, they gave him a shot with a needle and syringe combination that my memory says competed favorably in size with today's caulking guns. I swear! I remember it to be huge! And I was so thankful they were giving it to him — not me.

Now back to my bone marrow aspiration: Unbeknownst to the nurse practitioner and her helper, I had the image of a caulking gun and appropriately sized needle going through my mind while they were discussing the "short" versus the "lonnngggggg" needles. They were talking rather quietly too. Maybe they thought I couldn't hear them. Their words were close to a whisper in volume, but loud enough to hear. And when they said the word "lonnngggggg", they said it exactly the way I spelled it.

Maybe if I had seen the "needle" they were using, I would have been much more anxious, but they didn't show it to me. I only saw one after my third sampling. Just recently, I searched the net for it, and there are pictures there of the huge "needle" used in this procedure. Fortunately, they didn't show me (and I couldn't see) what they were doing.

A few minutes after their discussion about the needle, and after several warnings that "now you're going to feel a little pressure," one of them asked me, "Do you drink a lot of milk?" MM had weakened many areas in my bones, but it apparently had not done much to my pelvis — or at least the particular site at which they chose to take the sample. Their question indicated to me that they were having trouble pushing the needle into the bone to reach the marrow cavity.

When they finally reached the marrow cavity, they said I would feel a slight "pulling" sensation. Even with the local anesthetic, the withdrawal of some of the marrow (the "pulling") was painful — but it was over quickly — probably 10 seconds or less.

That was it! They cleaned and bandaged the site, and I was finished. Fortunately, the whole procedure didn't take very long (30-45 minutes), so I was out of there as soon as they allowed me to leave. There was no way I was going to hang around any longer than necessary — they might remember that they forgot to do something and ask me to stay longer!!

Second and Later Samplings

The second bone marrow sample was taken about two months later. It was performed at a different clinic by a different nurse practitioner and a different helper.

This time, before they started anything, I told my chemo nurse that I wanted a "happy" shot. At the time, I didn't know what I was requesting. I wasn't sure whether such a shot was even a possibility, but I figured I'd ask the chemo nurses using that vague terminology and see what they said. I also told them that if all else failed, they could knock me out completely for the procedure. As long as I wasn't subjected to a similar amount of pain as I experienced during the first sampling, I would be happy.

Putting the patient to sleep is standard procedure (according to the internet) before <u>collecting</u> bone marrow for a transplant. If they can do it before <u>collecting</u> bone marrow, they certainly can do it before <u>sampling</u> bone marrow. But standard procedure for sampling appears to be the way I have described — and they have yet to put me to sleep for it.

This time, after asking for the "happy" shot, the chemo nurse volunteered a morphine shot. They seemed surprised I hadn't received a similar shot during my first sampling.

Even with the morphine shot, my third and later samplings were all painful when they extracted the marrow. I thought maybe they gave me a smaller dose of morphine, but the nurse practitioner suggested the amount of pain was a function of the health of the bone and the individual. I only know that they have **all** been painful.

My overall experience with bone marrow sampling is that the procedures are not too bad. Using the morphine and local anesthetics, it is not a totally terrible experience. Having said that, it is still enough to keep me awake for long hours the night before the procedure.

To put this all into perspective — my Mother supposedly went through a bone marrow sampling procedure and after the doctor had finished and left the room, Mom asked my wife, "Do you know when he is going to start?"

That is my goal for bone marrow samplings — to be able to go through the whole procedure without feeling anything. It hasn't happened that way yet — and they promised me many more bone marrow tests. So there's always hope that a future procedure will be totally painless.

Options

The two main options that you have when told that you need a bone marrow sample are: (1) Where and by whom will it be performed? and (2) Are good pain medications administered prior to the procedure?

Regarding where and by whom, the doctors will follow their clinic's standard procedures. From my experience, the nurse practitioners at each clinic have been designated to perform these procedures. You can and should ask questions before the test and make your desires known. If you don't speak up, they will follow their standard procedures and make all decisions for you. After my first experience, I requested the procedure be performed at a different location by a different nurse practitioner. Now that I've undergone five or six such procedures, the staff knows whom I prefer to perform this procedure.

Regarding pain meds — before the first sampling, no one told me that I could request extra meds. Also, it was my first sampling, so I was entering unknown territory. I had no idea what was going to happen or how the test was performed. At the second and all later samplings, I received a shot of morphine along with the local anesthetic.

Be aware — you need a driver to chauffeur you around after this procedure. The pain meds affect your ability to drive — so don't even try driving alone to the clinic for this procedure. If they know you are alone, they may not perform the test at all, and they certainly won't give you a morphine shot. Secondly, having undergone several such procedures, I know that the last thing I want to do after being released is to drive the car. All I want to do is sit down in a comfortable chair and relax.

MRI – *Magnetic Resonance Imaging*

This procedure is not painful at all — expensive, but not painful. You must lie perfectly still for the whole procedure which can last a fairly long time (depends how much of the body they are imaging). In my case, because the MM damaged my spinal column, they wanted to see images of my whole spine. Each of my MRIs took 2½ hours to complete, and the pathology reports said "extensive" deterioration had occurred to my vertebrae.

Two of my vertebrae are crushed and the rest are deteriorated. My height after five cycles of chemo was 6'0", whereas for most of my pre-cancer life, I was 6'3".

MRI analyses do not cause any physical pain, but you must lie perfectly **still** for the duration of the test. The instrument is quite noisy, so they gave me ear plugs to wear during the procedure. Because the analysis chamber is small and tight, **claustrophobia** can be a major problem.

As far as medical tests go, there is nothing to fear about this one. They don't need to poke or prod or even touch you. The magnetic field allows analyses of any and all parts of the body — and you won't feel a thing. If you can sleep without moving, this is a good time to do so.

The Technique

The MRI (Magnetic Resonance Imaging) technique uses powerful electromagnetic fields which cause atoms and molecules in your body to vibrate. Our bodies are not magnetic, but under powerful magnetic fields, molecules do react — they vibrate — and each molecule vibrates differently. The instrument can detect these vibrations and use their intensity variations to create images.

To see three-dimensional images, they make many two-dimensional images of thin sections through the area to be analyzed. Then, with these images stacked like a deck of playing cards, they use their computers to produce the images perpendicular to the sections and to produce three-dimensional images.

Recent advances in precision in this technology produce very finely detailed images. Recent advances in computer speeds and computer software provide detailed images in relatively short periods of time.

The restriction on this technique is the presence of metal in your body. Some metals, like gold, are not attracted by magnetic fields. But iron alloys and a few other metals are magnetic and attracted (pulled with great force) by the strong magnetic fields. The technical staff will ask lots of questions before this procedure to learn whether or not you have any metal anywhere in your body. If you have any plates, pins, screws, etc., anywhere in your body, you are not a candidate for an MRI. If you have no metal implants, they can proceed. All external metals must then be removed from your clothing before the test. Pens, glasses, wrist watches, paper clips, metal eyelets, metal zippers, etc., must all be removed before entrance into the MRI lab because all are attracted by the strong magnetic fields.

Claustrophobia

Early models of the MRI instrument were large donut shaped machines. Patients lie on movable platforms that can roll the patient into the hole of the donut — the central cavity of the instrument. The opening is about two feet in diameter. Depending on your size, it can be a tight fit. I am fairly large and in the local unit, the inside top of the chamber appeared to be only about 4 inches above my nose. The inside length of the central tube is about two feet long and then it opens up again towards the rear of the instrument.

Once in this tube and in position for an analysis, there is no space to move around. I could move my hands a few inches from their starting positions by my sides but that was it. If your nose itches, there is not enough space to allow you to scratch it.

Those who are claustrophobic may not be able to endure an MRI analysis. To get around this problem, the staff can administer medications to help you sleep through the procedure.

Newer MRI models are C-shaped — that is, they are open to one side. On these units, it appears you are lying beside the instrument — not in it. The only tight spot is the central analysis section of the instrument, the opening of the C, which is open to the side and into the room.

If claustrophobia is a problem, the patient can always ask to have the test performed in a lab which has one of these newer MRI models or the patient can request to be put to sleep during the analysis.

In my case, I am not claustrophobic. Even so, since the chamber is really tight, I initially had to keep reminding myself that I was not claustrophobic. I shut my eyes for a while and that helped. Then, I realized that when I was pushed far enough into the chamber, my head was sticking out the back of the machine — so I began to look around to examine the back of the instrument room. That helped quite a lot. Once

I realized I could see out the back of the instrument, claustrophobia was no problem at all for me.

The Measurement Process

From the patient's point-of-view, the MRI measurement process simply requires you to lie perfectly still for the duration of the analysis. In my case, to analyze my whole spinal column required 2½ hours.

The MRI units they used for my tests were noisy. Magnetic fields aren't noisy, so I assume that the mechanical equipment required to focus the magnetic field and move the analysis from point to point produces the noise. There were lots of hums, beeps, and clunks throughout the procedure. This is not a problem, however, and it does provide something to think about as the analysis takes place.

Chemotherapy

The chemotherapy treatment for each type of cancer and for each individual will differ. Every treatment program will be specifically designed for the particular cancer and the particular individual being treated. In my case, I received the most aggressive treatment available for multiple myeloma. Since I had only little experience with cancer and with other cancer patients, I had no frame of reference against which I could compare the expression *"the aggressive treatment."*

My mother had a non-Hodgkin's lymphoma, and my father-in-law had brain cancer, but those were very different cancers from the one I have. They each received very different treatments than the chemo I received. They both had *the most aggressive treatments* available for their particular cancers, and I saw what happened to each of them. Mostly, their experiences made me wary and extremely cautious of X-ray treatments.

I allowed the doctors to treat my clavicular fracture with X-rays (to kill cancer cells residing at the fracture site), but I refused X-ray treatment of a "tumor" in a vertebra. First, it was my understanding that with multiple myeloma, one doesn't have "tumors" — one has lesions (holes). Secondly, they were wanting to irradiate my lower spine. I did not want to be a paraplegic. Thirdly, the doctor who suggested the irradiation was a neurologist, and cancer was not his specialty. So I said, "No!"

Now, post-chemotherapy, I still can only speak for myself and how I felt. Both parents seemed to do well as they underwent chemo treatments, so their experiences taught me that today's anti-nausea meds and chemo treatments are really good. Otherwise, I had little information and no experience on which to anticipate how well my chemo would go.

The official title of my program of chemotherapy was "Velcade—DT-PACE." The orders for this program are reproduced in Table 1.

My oncologist told me right away that this was the nastiest, most complex chemotherapy program that he gives to anyone he treats for any cancer.

There are more than half a dozen cancer meds in this program. There are also a bazillion other drugs that are administered to counter all of the possible side effects of the cancer drugs.

There are enough drugs in this program to play havoc with most possible side effects. One drug can take one side effect in one direction and another drug can take the same side effect in the opposite direction. This happens for most possible side effects.

For example, the main cancer drug – thalidomide – can produce constipation. Magnesium oxide and several other meds can cause diarrhea. Some meds can cause both. Other meds cause weakness, sluggishness, lack of appetite, and sleepiness while dexamethazone, prednisone, and other steroids rev you up.

Hopefully, all such side effects will achieve balance — but they all vary with each individual, the particular meds and dosages, and the combinations of all meds used. After my one year post-transplant checkup, my insides still were behaving badly. I was only then just beginning to feel like I was returning towards normal, and I had stopped several of the nasty meds at least two months before that checkup. One year post-transplant was about 18 months post-diagnosis, so my body had been messed up by the meds for 18 months or more and it still was not back to normal — but I was beginning to feel a little better, finally.

Then, there is the case where a drug can produce a major unwanted side-effect, and a second drug is prescribed to counter that particular side-effect. A major side-effect of thalidomide, for example, is the production of blood clots in your legs (deep vein thrombosis.) To counter this possibility, an anticoagulant was prescribed. Steroids send lots of sugar into the blood stream, so along with the concentrated steroids, another drug was simultaneously administered to kick excess

Table 1. Chemotherapy Orders

DTPACE + *VELCADE* Standing Orders

Diagnosis code__203.00__

*****This treatment regimen should be started on a <u>Friday (Day -2), Day 0 is Sunday.</u>****

Pre-medications;
1) Kytril 10mcg/kg = _2000_ mcg in 100ml NS IV over 10 minutes Prior to Velcade on Day -2 and Days 1-4
2) Decadron _10_ mg in 100ml NS IV over 10 min Prior to Velcade on Day -2 only
3) Aloxi 0.25mg IV on day 5 (do not give any additional 5-HT3 antagonists).

Chemotherapy Velcade
BORTEZOMIB (VELCADE) _1.3_ mg/m2 = _3.0_ mg slow IV push on:

Dates: __✔__ (Day -2 Prior to DTPACE) __✔__ (Day 1 of DTPACE) and __✔__ (Day 4 of DTPACE). At least <u>72 hours</u> should elapse between consecutive doses.

Chemotherapy: DTPACE (Days 1 through 4): Dates_____ – _____
- ➤ Cyclophosphamide 300mg/m2/day by continuous infusion = _750_ mg/day ✕4
- ➤ Etoposide 30mg/m2/day by continuous infusion = _75_ mg/day ✕4
- ➤ Cisplatin 10mg/m2/day by continuous infusion = _25_ mg/day ✕4

(Cisplatin, Cyclophosphamide and Etoposide are to be mixed together)
- ➤ Doxorubicin 10mg/m2/day by continuous infusion = _25_ mg/day ✕4
- ➤ Dexamethasone 40mg orally q day ✕4. **VERIFY THAT PATIENT HAS RX.**
- ➤ Thalidomide (target dose = 400 mg q day): start at 100mg q day, increasing by 50 mg q week if patient tolerating medication. Reminder: Coumadin 5 mg po QD starting dose (with Thalidomide)

Begin the following prophylactic medications on day 5 (VERIFY THAT PATIENT HAS RX):
- ➤ Acyclovir 400mg po BID
- ➤ Diflucan 400mg po q day
- ➤ Zantac 150mg po BID *or* Prevacid 30mg po q day
- ➤ Cipro 500mg po q 12 hours (adjust dose for renal insufficiency)

On day 6, begin (physician must check appropriate box):
- ☑ Neulasta 6 mg SC on day 6, OR
- ☐ Leukine® (GM-CSF) 250mcg/m2/day = _____ mcg/day SC until ANC> or = to 1500 for 2 consecutive days

Give the following IV fluids on Days 5, 7, 9 (check box for appropriate fluid). Outsource day 7 fluids, otherwise change fluid days to 5, 8, 10:
- ☑ _2_ liter(s) 0.9% sodium chloride IV over 1-2 hours.
- ☐ __ liter(s) 0.9% sodium chloride + 10 meq KCl/L and 2gm MgSO4/L IV over at least 1 hour per liter.

APPOINTMENTS:
Labs: BMP w/magnesium on days 1, 5, and 9.
Follow up appointments_____

_____   _____
MD Signature   **Date**
original to medical records/copy to pharmacy

sugar out of the blood stream. All of the chemo drugs break down the immune system, so a series of oral antibiotics were administered to fight infections that a compromised immune system couldn't fight. I was told that if I took the antibiotics, I **might not** get sick. If I **did not** take those particular antibiotics, I **would definitely** get sick.

Since most chemo drugs make patients nauseous, the first drugs they administered each day of chemotherapy were anti-nausea drugs.

The amazing part of taking meds is that you can take a whole handful of pills at the same time; they all go into your stomach; they are each processed individually by your body; and they each perform their own specific tasks **without interacting with each other.**

Doctors, and especially pharmacists, are supposed to keep track of the combinations of drugs being taken so they can look for possible drug interactions. In my case, all of the different drugs worked fine! (Medical science today is really quite amazing!)

PICC Line

In preparation for chemotherapy, the first procedure performed on me was the insertion of a PICC line (Peripherally Inserted Central Catheter) in the inside of my left upper arm. See Figure 1.

Although some people receive a semi-permanent port which can be accessed on the upper chest below the clavicle, I received a PICC line. Two lumens were exposed and available for easy access to my blood vessels throughout the six months of chemotherapy sessions.

The PICC line was anchored to the skin with a bracket backed by a strong adhesive. Two wings on the line (shown in the figure) snapped into the bracket. The bracket and the entry point through the skin into the vein were all covered with a dressing that contained a large, clear plastic window. The dressing and the anchor bracket were changed weekly.

The PICC line was used to draw blood out of my body and to infuse chemotherapy medications into it. Other than the minor inconvenience of the requirement that the dressing and exposed parts of the line remain dry at all times, the PICC line was quite convenient because it was unnecessary to stick my arm each day (venipuncture) to gain access to my blood stream.

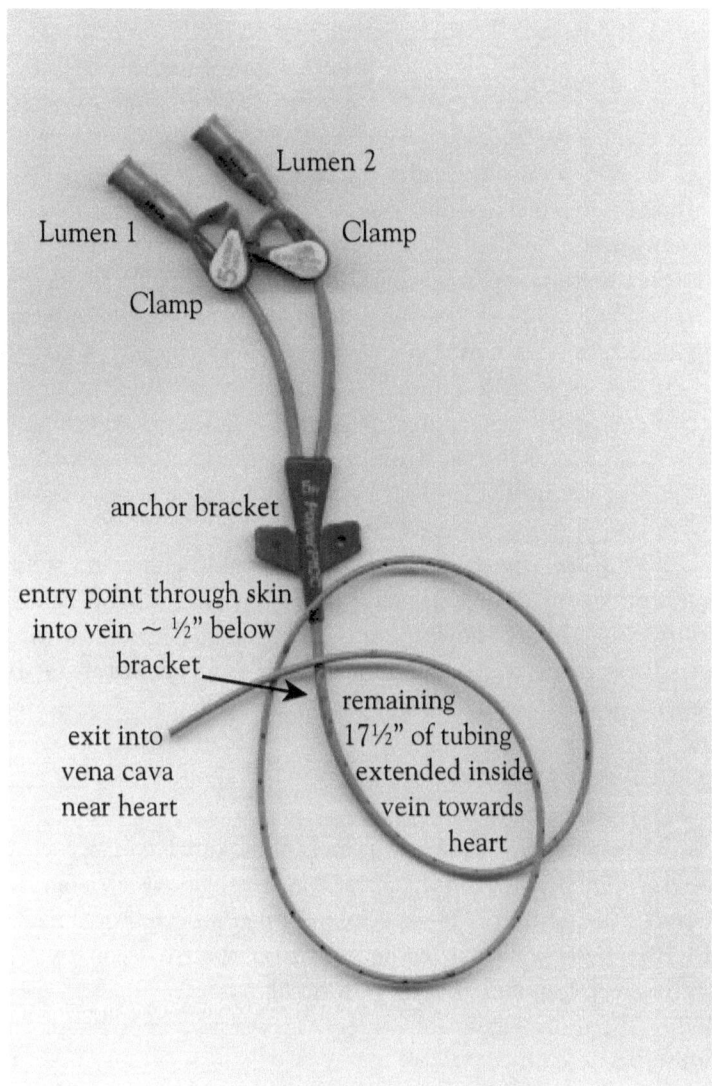

Lumen 2

Lumen 1   Clamp

Clamp

anchor bracket

entry point through skin
into vein ~ ½" below
bracket

remaining
17½" of tubing
extended inside
vein towards
heart

exit into
vena cava
near heart

Figure 1. PICC Line for Chemotherapy

Taking showers with the PICC line in place was tricky. We wrapped my upper arm with plastic wrap and we used packing tape to seal it to prevent shower water from accessing the tubes, dressing, or entry hole through my skin. To take a shower when pumps were attached 24 hours per day to the PICC line, it was necessary to run the tubes down my arm towards my hand and to hold the pump over the shower curtain rod where it could not be soaked by the spray. On those days, my wife assisted. Hanging onto the curtain rod helped me to keep my balance, but with the pump bag dangling over the curtain rod, it made the whole process awkward.

Fortunately, we have hand-held shower heads in all of our showers at home. That allows easy manipulation of the shower head and the spray with the free hand.

It was even more difficult to shower in the hotel room during stem cell transplant procedures because the hotel room had a fixed shower head. In this case, I had to hold the pump bag outside the shower while twisting and turning awkwardly to move around under the fixed spray.

To anyone who must go through chemotherapy and stem cell transplant procedures, I recommend that you install a hand-held spray nozzle in your primary shower if you don't already have one. If your current shower has a fixed-position head, hand-held nozzles are relatively inexpensive and easy to install.

A Typical Week of Chemo

The chemotherapy rooms at most clinics are large, airy, pleasant rooms with comfortable recliner chairs, blankets, pillows, IV stands, and IV pumps. Chemo nurses are specially trained and well qualified. All that I have met are nice, caring people who know what you need, how drugs will affect you, and how to perform all procedures. And they are sticklers — they don't stretch any rules. You will be in good hands during chemotherapy sessions.

In my case, treatments started on the day designated Day # -2, which was the Friday before the actual chemo week. This required that I present myself at the clinic so they could hook up an IV to one of my ports, give me the "pre-meds" which included anti-nausea drugs, and finally, inject the Velcade. Then, I sat around until the drip bag of saline was finished before they disconnected me and sent me home. Total time: about 2 hours.

On each chemotherapy day at the clinic, they drew blood samples and analyzed them while I was receiving the pre-meds. They were particularly interested in white, red, and platelet blood cell counts, as well as a hemoglobin number. Chemotherapy usually lowers blood counts, so they maintained a constant watch on the important values.

Depending what you are receiving through the IV, it could take anywhere from an hour to several hours to complete the procedure. Some meds, and especially transfusions, required slower flow rates which took more time. On those days, I just reclined comfortably, slept, read, watched TV, or visited with others around me.

On the Monday of each chemo week (Day #1), they connected a bag of saline to one of my ports so they could give me the dose of anti-nausea drugs and Velcade. Then, they connected two pumps containing appropriate cancer medications. Each pump worked 24/7 for the rest of the week to slowly and continuously pump the drugs into my system.

On Tuesdays (Day #2, and each day following through Friday – Days #3-5), they disconnected both pumps and connected a drip bag to my port to administer anti-nausea drugs. Thursdays (Day #4), they injected more Velcade into the line. Each day from Tuesday through Thursday (Days #2-4) they made sure there were sufficient medications in the supply bags for the two pumps. Then, when the bag of saline finished, they disconnected me from the drip bag, reconnected the two pumps, and sent me home.

On Fridays (Day #5), they disconnected the two pumps, gave me anti-nausea medications for the weekend, and followed with more saline. When they sent me home on Fridays of chemo weeks, I was free of the pumps and I didn't need to return to the clinic until Monday. At home each Saturday evening of chemo week (Day #6), I had to give myself a subcutaneous shot of Neulasta®. If I hadn't been able to do this myself, or if my wife or caregiver couldn't do it, I'd have had to go to the hospital to get the shot.

Mondays and Wednesdays of Week 2 (Days #8 & 10), I had to return for more IV saline fluids. If my blood counts were too low, they could give me blood transfusions. If my electrolytes were out of balance, they could give me whatever I needed (potassium or magnesium, for example) as well.

On Wednesdays of Week 2 (Day #10), they usually sent me home with directions to drink lots of fluids and call if there was a problem. It was a Thursday of Week 2 (Day #11, Cycle #2) when I felt terrible, and when they admitted me to the hospital with a fever and a slight infection. In most cases, I felt fine but weak after the chemo treatments. I usually saw the oncologist on Tuesday of Week 3 (Day #16) of each cycle of chemo. If I needed help at any time, I could call and get it. Otherwise, I was at home to recover.

All of my chemo treatments were performed as out-patient treatments. Only once did they admit me to the hospital — out of five cycles of chemo — and that was for a fever indicating a possible infection.

My blood counts were usually low towards the end of each Week 2. This was when I felt the weakest and when I was most susceptible to infections. It appeared to be a delayed reaction but I think the delay was typical — a few days after each chemo treatment was finished, the weakness, lethargy, and lack of energy hit. And then, in the weeks following, energy slowly returned.

Each Cycle is Different

After completing four cycles of "normal" chemotherapy, which in my case was the Velcade/DT-PACE treatments shown in Table 1, my first observation is that each chemo cycle and each response from my body had been slightly different.

The main oral cancer drug I was taking was stopped two days into Cycle #1. The IV chemo treatments and the oral treatment overlapped by two days.

During Cycle #2, both oral and IV meds were administered together. All chemo meds were entering my system at the same time.

During Cycle #3, the main oral cancer drug and the anticoagulant had been stopped in preparation for stem cell collection, so Cycle #3 consisted mainly of the IV medications.

During Cycle #4, the transplant doctor reduced the dosage of the main oral cancer med due to my increasing neuropathy problems. Both oral and IV meds were administered during Cycle #4, but the main oral med was taken at half strength.

The level of cancer in my body changed dramatically from its earliest detection through the end of Cycle #4. Before and during Cycle #1, I had 10-15% cancer cells in my bone marrow. After Cycle #2, there were less than 0.5%. So the drugs administered were reacting with the particular conditions found in my body at the moment. Cycles #1 and #2 had lots of cancer cells to fight. Cycles #3 and #4 have few cancer cells to fight.

I wondered how these meds affected my body if they could not find any cancer cells to attack? Good question. I guess the answer is "adversely." The physical effects of the four cycles appeared to be cumulative. I was much more tired after Cycle #4 than I remember being after Cycles #3, #2, or #1. In fact, the first cycle of chemo hardly affected me at all. At that time, I thought maybe the whole chemotherapy process was going to be a snap! But I learned otherwise. By the 4th cycle, I was weaker and more sluggish than I can ever remember.

I don't remember feeling particularly bloated after Cycle #1. Bloating was a problem after Cycle #2 when I was in the hospital, but not after Cycle #3. Then, again, it was a major problem after Cycle #4. The culprits, I believe, were the high concentrations of steroids which I took during each cycle of chemo.

Blood Pressure

My blood pressure was all over the map during those treatments. When I was diagnosed with cancer, I had high blood pressure and I was taking pills each day to lower it. But chemo also lowers BP. So I should have monitored my BP daily to know when to stop taking the BP pills. But I did not have a blood pressure cuff at that time, so it was only monitored on the days when I visited the clinic. As a result, I generally took my BP meds a day or two too many, so I learned how it feels when my BP is too low. I also learned what BP numbers define "too low." Now, I can tell by the way I feel whether or not my BP is too low.

I obtained a BP cuff at the start of the 5th cycle of chemo, which was the high dose/stem cell transplant week of chemo. Now, I can (and do) measure my BP daily (several times a day if necessary), so I can tell from the measurements if and when my BP is causing problems.

Blood pressure varies normally throughout the day, so that complicates matters. It is not simply a matter of measuring one's BP and having a definitive answer. The anxiety of going to a doctor's office can raise BP. The level of stress experienced at any given time can affect BP. The level of physical activity can affect BP. The level of bloating and fluid retention can affect BP. Tripping and falling can affect BP. Whether one is standing or sitting can affect BP. Etc. An isolated BP value here or there is insufficient to get a real handle on BP.

In my case, I usually stopped taking BP meds the week after chemo because my BP was OK without the pills. Then, when the chemo nurses thought my BP was getting too high, they directed me to start the BP meds again.

BP is a tricky balancing act – especially when you don't have your own BP cuff to measure it. On days when I visited the clinic, I knew what was happening — most days spent at home, I didn't know and I was just guessing. In such cases, I did my best, kept the telephone handy, and moved on. Now that I have a BP cuff, I can measure BP — which means, no more guessing.

High Dose Chemo & Stem Cell Transplant

Cycle #5 of chemotherapy was the high dose chemo followed by the stem cell transplant. During this week, they gave me high doses of the drug melphalan on Monday and Tuesday, fluids only on Wednesday, and my stem cells on Thursday. They did not give me any of the other chemo drugs or steroids during that week — only the high dose drug and anti-nausea meds.

I was expecting a relatively easy time during that cycle of chemo because the oncologist had suggested that the high dose chemo cycle was easier than the other cycles I had received. In some ways it was easier, but it sapped me of my strength more than any of the other cycles of chemo. In addition to my white blood cell count being close to zero, my strength was also close to zero. In that regard, it was worse than earlier treatments — although my extreme weakness could simply have been the cumulative effect of five chemo treatments.

Regarding BP reduction, the high dose cycle lowered my BP for a much longer time than previous cycles of chemo. For the first four cycles, I stopped taking the BP meds for maybe a week during each cycle. In the case of the high dose chemo, I stopped taking the BP meds for several months. My BP was good although my pulse rate was rather high during that time. The culprit, in this case, was a damaged heart muscle, which will be addressed separately in a later chapter.

Energy Levels & Blood Tests

Chemotherapy not only attacks cancer cells, but it also attacks good cells, compromises immune systems, and produces weakness and fatigue. Day-to-day changes, however, are subtle. There is no way to tell by feel how blood counts are doing each day. They need to be measured — and that requires a trip to the clinic.

When I complained of being extremely weak, one of the nurses questioned whether or not I was calling in at appropriate times when I felt bad. She suggested I may feel poorly due to especially low blood counts — and when that happens, they can give me blood products to solve the problem. My response was, "At what point do I feel so much worse than earlier in the day (or than the day before) that I need to make a call after hours to the on-call nurse?" She said I shouldn't sit around feeling totally weak — without calling. Since I felt totally weak almost every day since the start of chemo treatments, I asked at what point I should say, "I am **too** weak. This has gone far enough. I'm calling the clinic."? She couldn't answer that question.

Sure, if I can't stand up, or I can't walk at all — that is a change and a call is warranted. But if I am just incrementally weaker than earlier, is that worth an emergency call or not? It is not clear when the threshold is crossed. Each patient must pay close attention and decide this for themselves.

Side Effects

Weakness

The first and most noticeable problem I had with my chemotherapy treatments was weakness. The doctor predicted this would happen. He volunteered in one of our first meetings that he didn't expect me to get sick (nausea) during treatment, but he did expect me to be weak — weak to the point that I couldn't do very much but sit around. He was certainly correct.

I was somewhat weak after Cycle #1, and I couldn't imagine how I could get much weaker. But Cycles #2, #3, and #4 were each progressively worse. The high dose Cycle #5 was worse yet. Now, I wonder how much weaker one could actually be and still be able to function normally. Not much weaker, I think!

Legs

Other than overall weakness, the location of my most prominent weakness was in my legs — specifically in my thighs. Standing and sitting were OK, but the process of standing up, sitting down, or going stairs requires thigh muscles. Those muscles were definitely weak. I used to be able to run up and down stairs without giving it much thought. After Cycle #4, I was going up and down stairs very slowly, one step at a time. Chris said I was puffing after coming up the stairs one night after Cycle #4. That night, I was dealing with bloating; it felt like I had 20lb weights strapped to each ankle; and it was a chore to go up the stairs.

In the mornings, I was generally at my peppiest. But as each day wore on, it became more and more difficult to stand up and go stairs.

After the 4th cycle, I concluded that I couldn't get much weaker and still be able to walk. The next step would be to use the wheelchair to get around. After the high dose cycle, I learned that you **could** be weaker and still walk, but the wheelchair was definitely welcome.

In addition to the fact that my bones were weak from the MM, I twisted my ankle one day after Cycle #3. That ankle healed OK, but it felt like it could overturn easily again. Following that event, I gave walking much more care and attention. I wore shoes whenever walking around because they provided grip. With every step, I was aware that my ankle could easily turn again.

It is a weird feeling while walking on a flat surface to know that you can easily turn your ankle. With weak legs, it would have been easy to relax the muscles and let the ankle turn again. That could not only further injure the ankle, but the ensuing fall could cause other injuries as well as broken bones.

I used a cane since starting chemo Cycle #1, and after Cycle #4, I sometimes used two canes. They helped maintain my balance, but they also allowed better distribution of the forces required to stand up and move around. To go stairs or to stand up from a chair felt like I was applying as much force through my arms to the canes, or to the stairway's hand rail and a cane, as I was through my legs.

Comparing these requirements to my remembrance of how easy it was to run up and down stairs before cancer shows how really weak I had become.

Walking, following Cycle #1, wasn't too difficult. But as the cycles progressed, walking became more and more difficult. I learned from experience (unfortunately) that I needed to keep my knees locked to maneuver properly. I fell at the bottom of our stairs one day because I tried to take an extra step (the last one) without having proper hand support and without having my knees locked. My first leg hit the floor and I put weight on it before the knee was locked. I didn't have the strength to pull it into the locked position. So it began to buckle. Then, my second leg hit the floor and it did the same thing. The result was that both knees buckled and I went down onto the carpet, landing on both knees.

Fortunately, it was a soft carpet. I put a nice brush burn on each knee, however. And since my blood counts were low, the brush burns didn't heal for quite some time. I was thankful I didn't break any bones. The brush burns and later scabs on both knees became a constant topic of conversation whenever I visited the clinic wearing short pants, which was most of the time. The nurses couldn't walk past me without asking what I had done to my knees.

After that, I made a point of locking my knees before each step. It was clear that it would be relatively easy, while walking, to lose control of my locked-knee position and have my knees buckle again. Just putting weight on each leg was enough to cause the feeling that the knee could easily buckle at any time, were I not completely careful about it.

That's the same feeling I had in my right ankle. So I had to be doubly cautious. It would be easy to turn my right ankle or unlock my knee and fall again, even though I was walking on a flat floor. With the weakened state of my bones from MM, I didn't want to do either. So I learned to walk extremely slowly and carefully to prevent such events.

The consequences of a fall were much worse than the extra effort required to walk slowly and deliberately.

It is a totally new experience to need to be careful when walking slowly on flat dry surfaces because one's knees might buckle and one's ankles might turn. It isn't something I ever gave a second thought to throughout most of my life, but post-chemo, I needed to pay attention.

Arms

To a lesser extent, the weakness also showed itself in my arms. I didn't need to use them very often in a strenuous way, so I was not constantly reminded that my arms were weak, too. But every time I needed to stand up from a chair, or sit down in a chair, I was reminded.

It was relatively easy to free fall into a cushioned chair — as long as I was positioned properly before I did so. To sit down slowly onto a hard chair with or without arms was much more difficult. The cane and a table, counter, or chair arm usually provided the two points of support needed. When those two points of support were not available, it was much more difficult to slowly sit down and slowly get up again.

I learned that to use my arms like this also required more forethought than one would expect. To slide my rear end around to reposition it in a bed or other soft, cushioned chair required the use of my arms. This had always been a process that took no thought. Jamb the arms into the cushion, unload the weight from your rear end, and slide to a new position. Putting this much stress on my wrists, however, was a problem because I worried about breaking bones if my arms and hands weren't positioned properly. Moving around in a cushioned chair or bed, therefore, required forethought to make sure my wrists were in a position that could handle the physical stress.

I found that just putting my hands flat by my sides was a good anchor position, but if I lifted my weight with my wrists in that position, I was torquing my wrists. Making fists and jamming them down into the cushions worked better, but the wrists can still be unnecessarily torqued in that position, too. I don't have a good recommendation for this, other than to be careful and think about what you are doing. It is so easy for accidents to happen when your mind is out of gear — you need to be thinking about what you are doing at all times. Taking the mind out of gear and behaving as if you are normal, and as if nothing has changed, can cause lots of accidents.

Think about every little thing, and think often.

Balance

Another area of noticeable change was to my balance. I don't know whether this is characteristic of all who undergo chemotherapy, or all who suffer cancer, but I assume it is the result of both.

When I originally started using a cane, it was not for walking, but for standing. My balance was shaky whenever I tried to stand still and read labels in a grocery store or talk to someone. The cane provided the third leg of the tripod and with it, I was fairly stable.

After four cycles of chemo, I needed the cane to help me walk. And even with the cane, my free hand was constantly reaching out and finding another point of support. Walking down the hallway, I would lean my free hand on the wall or grab door frames as I passed. As soon as a table or cabinet appeared, I would lean on it.

Putting on my pants had never been a balance problem before. But after chemo, to stand (and balance) on one leg while putting the other leg into the pants, became a major problem. This became progressively worse with each cycle of chemotherapy. After four cycles, it was necessary to sit on the edge of the bed or a chair, or to hold onto a sturdy support like the bed post to accomplish this task. You can't fall when you are seated, so that became the method of choice.

Taking a shower also presented a major balance problem. Trying to stand still and balance with eyes closed was essentially impossible. Just trying to stand still in a shower with eyes open was a problem. If your shower has a good sturdy bar to hang onto, that is good. If a strong support is not available, leaning against the wall may help.

As I mentioned earlier, showering is tricky when you are dealing with chemo pumps which put meds into your system 24/7. Even without the chemo pumps, balance problems remained. During those days, it was easier (and safer) to take a shower when someone was standing by to help if needed.

Certainly, it is necessary to have a good slip resistant mat in the bottom of the tub. After many years without such a mat in our shower, it only took one slip on the soapy tub to cause me to go down (albeit in a somewhat controlled, slow manner.) One minute I was standing, and the next minute, I was sitting with my leg and knee wedged between the tub sides. I didn't hurt myself, but I could have. We purchased a good mat later that very day.

Standing up from a chair also presented a balance problem. I started public school with a childhood friend who came down with polio. When he returned to class after several years, he had braces on both legs. I remember when he wanted to stand up, he had to push himself into a standing position and then reach down and lock each knee joint in his braces before he could walk away.

With all of my balance problems, I needed to think and stand up slowly like that, too. If I didn't lean far enough forward when I stood up, I could easily fall back into the chair again. If the chair happened to slide away when I stood up, I could fall back, miss the chair, and end up on the floor. I did that once — everyone heard the loud thump on the floor and came running. Fortunately, when they arrived, they found me lying on the floor laughing. That incident, however, taught me that it was necessary to lean front far enough to properly center my weight before I could stand up.

Who thinks about such things? We want to stand up — we stand up. Period. That's it! But following chemo and with balance problems, one has to think about all sorts of little details that never required any thought before.

Diarrhea-Nausea

Immediately after diagnosis, the doctor told me that he didn't expect me to get sick during or after chemo. He was referring to nausea. I had both diarrhea and nausea one day which occurred because I ate too much the night before (I think.) I didn't like it, so after that, I became much more careful about what I ate and how much I ate (and it didn't happen again).

The doctor said they have really excellent anti-nausea drugs today, so he didn't expect any such problems during my treatments. I remember my mother saying that very thing when she went through chemotherapy nine years earlier. She never became sick during her treatments, and other than that one day, neither did I.

I know that some people get nauseous easily and chemo treatments can do that to you. I don't know the cause in their cases. I do know, however, that today's anti-nausea meds **are** very good!

Chemo affects each individual differently, and whether one experiences bouts of nausea depends on their own particular body constitution and how they react to the chemo meds. It is certainly not something that I worried about ahead of time. The first bag of IV they gave me every day during chemo weeks contained anti-nausea meds. I also carried a bottle of anti-nausea pills which I could use at any time as needed.

During high dose chemo, I was attached to a pump 24/7 to dispense anti-nausea meds. It was very comforting to know that if I had gotten nauseous, I could press the button on the pump and put 2cc of the anti-nausea meds directly into my blood stream. Taking pills when your stomach doesn't want to hold anything down can be a problem, but that pump put the meds directly into my blood stream where they were needed. I never actually needed to push the button for that medication, but like I said, it was very soothing to know that if I felt the slightest bit nauseous, those meds could easily be dispensed to immediately control the problem.

Neuropathy

The most intense side effect I had to deal with was peripheral neuropathy, which occurred due to several of the cancer drugs.

The nerves in fingers and toes, starting at the tips and working back toward the body, can be damaged by chemo drugs. This causes tingling and numbness. Fortunately I did not have a severe case, but it became worse with each cycle of chemo and with the maintenance meds post-chemo.

The doctor asked me one day, "Are you having a problem with neuropathy?" I answered, "Yes." He started to turn and enter my answer into the computer when he stopped, turned back to face me, and ask, "How severe it is? Can you feel the floor under you when you walk?"

He then explained that in severe cases, some people's legs go numb so they can't feel the floor when they try to walk. I answered, "Yes, I can feel the floor when I walk. Apparently, therefore, I only have a mild case." His dictation from that office visit indicates that I had a mild case of peripheral neuropathy (PN).

PN begins by feeling like carpal tunnel syndrome. Most of my friends have had carpal tunnel problems in their hands, so I imagine most people know how that feels. The tips of the fingers and toes are tingly like there are a bazillion pins sticking into them. The tingly feelings come and go for no apparent reason. One day, fingers and toes will feel terrible, and the next day they feel fine. After a while, they mostly feel terrible.

As PN develops, however, the sensations move up the fingers and toes toward the body. After the first cycles of chemo, the tingly feelings were limited to the tips of my fingers and toes. After four cycles of chemo, the tingly feelings had encompassed a joint or two on each finger and had moved in a corresponding manner from my toes to my feet. At that point, my fingers and hands were tingly and they hurt almost constantly. My feet and legs felt similar. One month post-transplant, my hands hurt — especially when handling something cold, like a can of soda. During those days, I purchased and wore arthritis gloves. They helped a little. When handling cold items, they helped a lot.

One of my meds was specifically for neuropathy pain, but it didn't work very quickly. Nerves don't heal quickly (if they heal at all.) The pain killers that I was taking as-needed didn't help with the neuropathy pain either.

At one year post-transplant, the neuropathy problems hadn't yet gone away. It was more manageable in my hands, but it had worsened in my feet. If this is the price I must pay to eliminate the cancer from my body, however, I'll take it — (but I wish it would go away.)

Taste

It is common, and should be expected, that your sense of taste will be affected by chemotherapy. At some points in my treatment, everything tasted like my mouth was adding two bags of sweetener to every bite I took. Foods and drinks that were normally bitter not only became sweet, but too sweet. Diet sodas were too sweet. Sweet tea was awful! Unsweet tea was OK because my taster automatically sweetened the drink. Water was sweet. Most foods were too sweet.

In a search for something to drink that was not too sweet, I discovered that unsweetened lemonade, with its tartness, canceled the extra sweetness added by my mouth and still tasted somewhat tart. Everything else tasted bad.

Then, there was also the metallic taste which simply made all food and drink taste bad. It was like drinking, eating, and chewing with iron nails in your mouth to add flavor. There was no solution to this problem. It was simply a matter of suffering through the bad metallic taste until the effects disappeared.

Changes in your sense of taste, of course, depend on the particular chemotherapy meds used and the interactions between that regimen and your body. If a person undergoes chemotherapy and never experiences taste changes, they should consider themselves fortunate.

When and if taste problems arise, simply experiment with different food and drink combinations until you find something that tastes good.

Hair Loss

Most cancer drugs affect body cells that grow and reproduce quickly. This primarily targets cancer cells but also affects fingernails, hair, and skin. Usually, chemotherapy patients **will** lose their hair. In my case, the hair on top of my head grew thin and eventually looked and felt like peach fuzz. It remained that way for a long time during and following the five cycles of chemo. The hair on top of my head and my mustache were my most noticeable hair losses. Most of my kids had never seen me without a mustache and certainly never without hair. My hair is normally brown, but the fuzz on top of my head was mostly white — actually, it was mostly invisible.

I shaved my mustache soon after chemo began. One day, I washed my face with a wash cloth and rubbed half of my mustache off — but I didn't notice it in the mirror. It didn't feel good, so I didn't rub the other half with the same vigor. Then, out in public, my wife commented that I only had half a mustache. I fixed that ASAP.

A lot of the hair on my body disappeared without notice. The hair on my arms and legs, however, looked normal throughout most of the process. Well after the high dose treatment when the hair on my head finally began to return, the remaining hairs on my arms and legs fell out.

I was told that following the "high dose" cycle I would definitely lose all my hair, but they had told me that same thing after all four regular chemotherapy cycles as well. I interpreted that to mean that the fuzz on top of my head would disappear, too, but it never actually happened.

One clinic staff member, in an attempt to encourage me, looked at my fuzzy head one day and said, "Oh, I see your hair is starting to grow back!" Unfortunately, it hadn't fallen out yet — I was still waiting.

Hair loss goes with the territory. You get chemotherapy — you lose your hair. Some meds are stronger than others, and some body constitutions are stronger than others. So once again, it depends on the individual, the specific chemo treatments used, and the strength of one's body constitution. Expect to lose your hair and be pleasantly surprised if you do not!

My facial hair was the first to grow back following transplant. The hair on my head began to grow again around 100 days post-transplant.

Heart Problems

I had not read about nor anticipated heart problems as part of my chemo treatment. I had them, though, as a direct effect of my chemotherapy. The subject is important enough that I gave it a separate chapter in this book. It is amazing to me that I had not read anything about this prior to completion of my chemo treatments. When I specifically went looking for mention of such possible problems, I found it.

Read more about this in a later chapter.

Direct Effect – Kill the Cancer

There are several direct effects of chemotherapy that can be expected to occur with chemo regimens.

Of course, the obvious primary effect of chemotherapy is the elimination of cancer cells from one's body. In my case, the treatments appear to have accomplished this very well.

Decisions regarding which particular chemotherapy regimen to use are between doctor and patient (the doctor analyzes, tests, decides, and recommends, the patient asks questions, seeks advice, and approves or disapproves.) Treatments vary with particular cancers, but be assured that in the early years of the 21st Century, cancer research is moving along very quickly and successfully. Oncologists have access to the latest test results and best treatment methods, so they know which chemotherapy programs are best for treating each different type of cancer.

My Treatment

In my case, the first two cycles of chemo effectively removed all but a trace of the cancer from my system. Cycles #3, #4, and "high dose" were used for purposes of being **thorough** — that is, to raise chances for a long period of remission following the treatments.

I specifically asked the transplant doctor about this very point. If the level of cancer in my body was essentially zero after two cycles of chemotherapy, why should I need to go through a 3rd, 4th, and 5th cycle? The result of my first bone marrow biopsy, which confirmed the cancer diagnosis, showed 10-15% cancer cells in my bone marrow. The result of my second bone marrow biopsy after Cycle #2 showed less than 0.5% cancer cells. The doctors said this was effectively a zero. The answer to my question was that it may be easy to eliminate the cancer in bone marrow, but this cancer can also reside in the bones. Cancer cells here or there in bone are less easy to detect, analyze, reach with the meds, and eliminate. So to make sure we had done our best to reach those cancer cells, the 3rd and 4th cycles were given prior to the 5th high dose cycle. Studies have shown that length of remission increases after 4 cycles of the aggressive chemotherapy, compared to patients only treated with 3 or 2 cycles.

So the doctor's answer was a good answer. If going through more, nasty cycles of chemo during the first treatment pass means I may not need to go through the whole process again **later** — that is good! I can handle it.

Having been diagnosed with multiple myeloma, I was fortunate at 60 years of age to be considered "young and healthy" which qualified me for the aggressive chemo treatment and the stem cell transplant. This treatment, known as the "Arkansas" treatment, was the best available for multiple myeloma in 2008.

Secondly, I was fortunate that the Cancer Centers of the Carolinas and St. Francis Hospital, both of which are in Greenville, SC, offer this treatment package on an outpatient basis.

We learned that a friend of a friend was also receiving this same treatment, but she was doing it with the doctor in Arkansas. This meant she had to travel out there frequently to complete the process. She was diagnosed 5 years ago and at that time had little option but to travel there to obtain the treatment. The Transplant Director, who is in charge of my treatment, is a colleague of her doctor and he treats his patients here in the Upstate of South Carolina.

I also learned of another woman whose health insurance refused to allow her to be treated in Greenville. To meet her health insurance's requirements, she had to travel elsewhere.

I was fortunate that our health insurance company recognized and supported this aggressive treatment process here in South Carolina. At the time of my diagnosis, there were only two locations in South Carolina that were approved by our insurance carrier for this aggressive, chemo treatment/stem cell transplant process. One was a hospital in Charleston and the other was the team consisting of the Cancer Centers of the Carolinas and St. Francis Hospital in Greenville.

All of the treatments I received were performed as outpatient procedures. By 2008, treatment methods had advanced to the point that they could offer the aggressive treatment this way — and I happened to live in an area where it was available.

They only admitted me to the hospital once for five days after Cycle #2 of chemo because I had a fever — which indicated that I had an infection. The doctor said it was "urgent" that I be treated for that infection quickly — and I was. The critical day of treatment for that infection was that first night in the hospital. I was admitted on a Thursday. The urgency had passed by the time I saw the doctor Friday morning. That was good to hear.

Apart from that one hospitalization, all chemo treatments and the rest of the procedures, including stem cell mobilization and collection, high dose chemo, and stem cell transplant were all performed at one of the Cancer Centers or in the outpatient wing of the hospital. No other inpatient hospitalizations were needed.

This saved a lot of money because the average daily cost for each of my five days in the hospital was about $5000. As an outpatient, I lived in a hotel near the clinic for two weeks during the high dose chemo and stem cell transplant procedures. Even if it had been an expensive hotel (which it was not), two weeks in the hotel were much less expensive than two weeks in the Bone Marrow Transplant Ward of the hospital.

Another advantage, according to several nurses and doctors, is that hospitals are a great place to pick up infections, simply because there are lots of sick people who visit hospitals. Housing me at a local hotel during transplant procedures at the clinic helped prevent my exposure to all sorts of bugs, viruses, etc., that could lead to infections. During those days, I was either in a small, separate room at the clinic, in my hotel room, or traveling from one to the other. I wore face masks when my immunity was weakest and when I needed to walk through common areas at the clinic. Otherwise, I spent many hours in my room by myself where I was not exposed to anyone but one or two chemo nurses and my care giver.

The Lord's Control

Unknown to me at the time of diagnosis (or prior to that), the Lord had positioned me well to best attack this cancer:

(1) I live in a home where everything necessary is available on the main floor. I hadn't ever thought about this, but a friend pointed it out to me one day when discussing how weak these treatments made me. It certainly was beneficial to not need to go stairs several times a day.

(2) It was not clear why the Lord appeared to have my wife working a less-than-wonderful job. We had discussed this point several times prior to diagnosis, but the answer came almost concurrent with the diagnosis. Chris' job provided me with excellent health insurance which covered my whole treatment process at minimum expense to us.

(3) The oncologists, to whom I had been referred by my Christian family doctor, are an outstanding team of doctors.

(4) The treatment program that I was able to receive locally was state-of-the-art.

And best of all, (5) the fact that they considered me to be "young and healthy" qualified me for the aggressive treatment, and throughout those treatments, my body cooperated well with all medications prescribed.

Point (1) had not entered our thoughts or considerations before this time. We weren't living in our home because it has kitchen, bedrooms, living room, and bathrooms all on the same floor. At the time we purchased this house, the Lord appeared to have provided an ideal house for our family. We hadn't consider the idea that if one of us became sick, we could live on one floor and never need to use the steps or the lower level. We never even considered the idea that one of us could become sick. But the Lord knew ... and we moved into this house 20 years ago.

Regarding Point (2), we had often wondered why the Lord wanted my wife working at her food service job — or working at all. We prayed about it and she applied for quite a few jobs before finding this one. We now believe the Lord put her in a job which provided excellent health insurance coverage. At the time, we didn't know what was happening, or what was about to happen, but the Lord knew!

Cancer was not something we ever thought about or made plans for. The most we had done was purchase a supplemental cancer insurance policy, and once that was done — it, too, was out of sight, out of mind. Regarding Points (3) and (4), we had no knowledge of cancer, oncologists, cancer treatment centers, or cancer treatment programs. All such subjects and thoughts were totally off our radar.

Following my diagnosis, I learned that on the drive to and from my teaching position at OCA every day, I passed within a few hundred yards of the Cancer Centers of the Carolinas Seneca Clinic. My doctor works in both the Seneca and the Greenville Clinics. In conjunction with St. Francis Hospital (a few miles from the clinic in Greenville), the Greenville Clinic offers state-of-the-art treatment programs, i.e., stem cell transplants. I didn't have any idea that we lived that close to such outstanding capabilities. But the Lord knew all of that!

Finally, point (5) deals with my body cooperating with the medications. I have no control at all over this one. At 60, I certainly didn't consider myself to be "young," although I was definitely healthy for the first 60 years of my life. I had no idea that those two characteristics would qualify me for an aggressive treatment for this cancer. I also had no control over how my body would react to any of the medications they gave me. I received frequent compliments by the medical staff on how well I was responding to the meds. I had nothing to do with it.

Concentrating with all of my might on trying to make my body respond well would have had no effect at all on how my body actually did respond.

Matthew 6:27 applies: **"Which of you by taking thought can add one cubit unto his stature?"**

We can't change our stature by thinking about it, any more than we can cause our bodies to cooperate well with meds. It was totally out of my control. It has always been **out of my control**; it remained so when I was diagnosed with cancer; and it remains so today. But it has been **well within the Lord's control**! He saw to it that my body responded well to the medications. He knew what was happening to me and He made it all work well.

When I look back, I am amazed that I didn't make any stupid decisions along the way and mess everything up. During the several years prior to my cancer diagnosis, we were trying to be patient and to wait on the Lord — but that is easier said than done. During that time, I either didn't make any stupid decisions, or more likely, I did make some stupid decisions and the Lord worked around them to keep us in line with His plans for us. Finally, the Lord helped us to remain patient through all of this.

The Lord is and has been in control of our lives and in control of my particular cancer situation. Unbeknownst to me, He put us in the best possible location and situations for treatment of my cancer. My family and I continue to be blessed by Him as I fight this cancer.

10

Autologous Stem Cell Transplant – The Collection Process

The terminology for this procedure needs some explanation. In the recent past, what I received would have been called a *bone marrow transplant*. In those days (as I understand it), they would find a donor whose bone marrow matched the patient's; collect bone marrow from the donor; collect blood stem cells from the bone marrow; and transfuse them into the patient at the appropriate time.

It appears to be the case, both then and now, that the important cells for this procedure are blood stem cells. So the procedure should really be called a *stem cell transplant*.

Today, they know (1) how to encourage the multiplication of your body's own blood stem cells, (2) how to encourage those stem cells to move out of the bone marrow and into the circulating blood stream, AND (3) how to easily collect them. Note that this can all be done within a single individual — that is, within the patient! It is still an option to try to find a match to collect and use a donor's cells, but your own stem cells are the best possible match to your body!

The word *autologous* is used to show that the transplant is happening within a single individual — that is, within the patient. The patient receives several cycles of chemotherapy. Then, using special medications, blood stem cells are encouraged to multiply and enter the blood stream. When stem cell concentrations in the blood stream are sufficiently high, they are collected, treated, and stored. Following stem cell collection, the patient is given the remaining chemotherapy treatments. Finally, the patient is given a cycle of high dose chemotherapy followed by the infusion of their own stem cells back into their bodies.

The point to be emphasized here is that the stem cells which were removed from the patient's body and stored at the blood bank are not present in the body during the high dose cycle of chemotherapy. Because the stem cells were removed from the body, they were not subjected to damage by the high dose meds. When the high dose meds are no longer capable of damaging good cells, the patient's stem cells can be re-infused by IV so they can restart the body's normal functions to create and produce new, good blood cells.

A quick summary of the stem cell transplant process is: cause stem cells to multiply and mobilize into the blood stream; collect, remove, and protect the stem cells; finish chemotherapy treatments to eliminate cancer cells; and then return the stem cells to the body where they can again function normally. Using this procedure, the whole body (except for the collected, protected stem cells) is subjected to the final high dose of cancer medications. Since the stem cells were not subjected to the high dose medications, they could not be harmed in any way by those meds. Upon their return to the body, they can once again carry on their assigned tasks.

The overall procedure is really slick — and today's medical science, understanding, and instrumentation have made this happen! Past treatment processes were a different story; but we are here beyond 2008 and today's capabilities are **absolutely amazing**!

The Collection Procedure

The whole stem cell transplant procedure sounds scary and complicated, but from personal experience, it is straight-forward, simple, and not scary at all.

The central instrument that makes this whole treatment possible is a stem cell collection device based on a continuous centrifuge. When whole blood is spun in a centrifuge, relative to the other blood cells, stem cells float because they are light. This allows stem cells to be separated from the rest of the blood, so they can be collected and stored.

The collection instrument contains lots of tubes, pumps, valves, storage bags, and the continuous centrifuge. Through an IV tube, blood is removed from the patient to flow through this machine, stem cells are separated by the continuous centrifuge, and the remaining blood is remixed, warmed, and returned to the patient almost immediately.

My collection day was long, lazy, pleasant, and comfortable. Once I was connected to the machine, I just had to sit around for six hours until the collection process was finished. The collection process can continue for up to five six-hour days until the desired number of stem cells is collected. Connection and disconnection from the machine is simple, so this process represents up to a week of lounging around and waiting to collect the appropriate number of stem cells.

In my case, this was a relaxing, lazy process. I had brought lots of things in my carry bag to keep me occupied. I could work crossword puzzles, work on crafts, read, write, work on the computer, listen to the radio, watch TV, watch movies, eat, drink, sleep, etc. I was restricted to the bed next to the collection machine, but that was the only restriction.

Since this process varies with the individual, it could take several days, and it could be less comfortable. Medications were administered to me before and during the collection process to prevent problems. But everybody is different.

My experience was a good one. I was relaxed and comfortable, and the collection process was completed before I knew it! I had a good experience. Others have not. It depends. Fundamentally, this is not a terrible procedure and it is not a process to be feared.

Central Venous Catheter

The scariest part of the whole stem cell collection process centered around the "central venous catheter." (See Figure 2.)

When the oncologist originally gave my wife and I a brief explanation of the collection process, he said that they would take blood from my body through an IV tube (as he gestured toward one of his forearms), put it through a machine, collect the stem cells, and return the blood through another IV tube (as he gestured toward his other forearm.) That didn't sound bad at all.

The information the transplant coordinator gave me, however, suggested that they would install a large catheter in a vein near the base of my neck to accomplish this task. That sounded much nastier! And I was NOT looking forward to it.

I already had a relatively permanent (but nevertheless temporary) catheter (a PICC line) in my left arm. They said they couldn't use that because its tube diameters were too small. They needed more blood flow to collect stem cells. So the catheter they put into my neck had larger diameter tubes capable of greater blood flow. The principle sounded OK but the fact that they were going to place it in my jugular vein didn't sound good at all.

I knew that on the day they wanted to start collecting my stem cells, they would first need to insert the new catheter into my neck. I was more anxious about the catheter insertion procedure than about the collection of the stem cells.

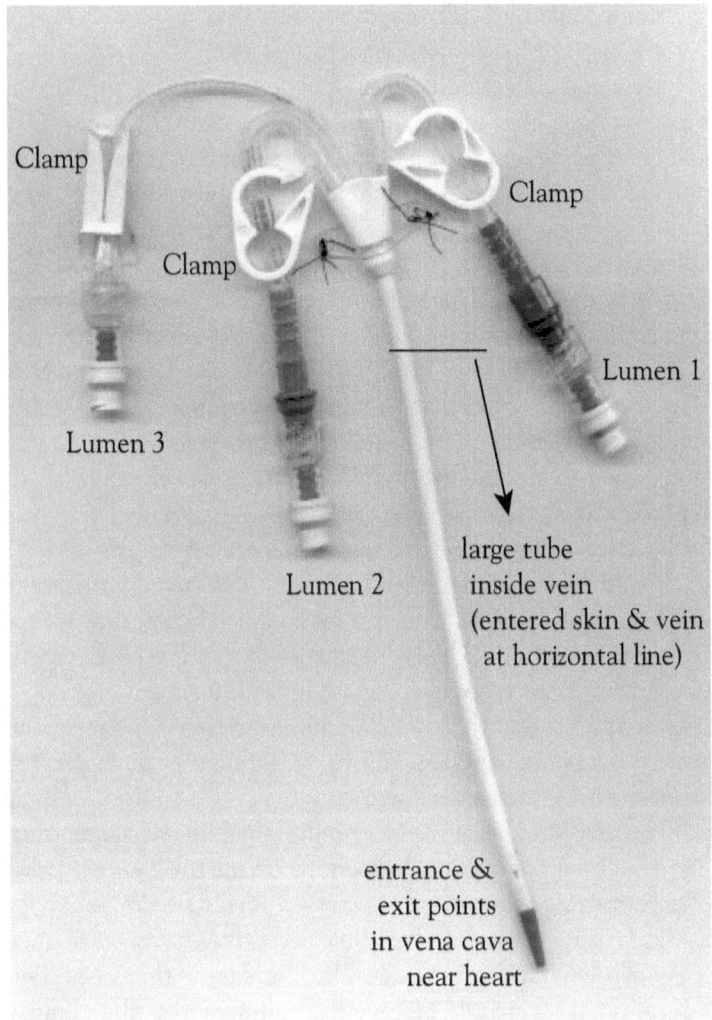

Figure 2. Central Venous Catheter for Stem Cell Collection

Again, **my anxieties were wasted.**   The installation of the catheter was another smooth, uneventful process that took all of ten minutes to complete.

Requirements & Timing for Collection

Stem cell collection can only happen when certain blood count requirements have been met. This means they can tentatively schedule the collection procedure for a particular week and day, but they need to measure blood counts and determine the number of stem cells per cubic centimeter of blood before they can actually determine the day to start the collections.

In my case, they blocked out a two week window following the week of my third cycle of chemotherapy. Chemo was administered in the Seneca Clinic from Monday through Friday as usual. On the Monday following the week of chemo, I needed to report to the clinic in Greenville so they could check my blood and give me Neupogen shots to encourage the production and mobilization of stem cells. I learned that they had tentatively scheduled me to collect stem cells on Tuesday of the second week following chemo — but it all depended on my blood counts.

Every day, starting on that Monday and lasting until stem cell collection was finished, I was given Neupogen shots. The Neupogen dosage is proportional to body weight — so in my case, each dose consisted of two 4 ml syringes to be administered subcutaneously. The needles were the same as those used for administering insulin to diabetics; they were thin, pointy, and easy to use; and I was able to give myself the subcutaneous injections each evening.

The first Monday at the Greenville clinic, they demonstrated the subcutaneous injection procedure as they gave me the Neupogen for that day. Each evening following, I was to self-inject two more Neupogen shots. Each time I visited the clinic after that, they gave me new Neupogen syringes for that evening. On Thursday of that week, I moved into a local hotel so I would be only five minutes from the clinic. The hotel was conveniently located a few blocks away. My hotel reservation was open-ended because they didn't know how long it would take to complete the stem cell collection process.

A normal white blood cell (WBC) count is supposed to be 4500/ml or higher. After Cycle #2 of chemo, my white blood cell count had fallen to 200. This time, on Friday of the week following chemo Cycle #3, my white blood cell count was down to 400. The WBC was low, but not quite as low as after Cycle #2. The nurses pointed out numbers on the blood analysis results that showed that the Neupogen was working and that the white count would soon rise — but by that Friday, nothing had yet happened.

To collect stem cells, the WBC had to be above 5000/ml and the platelet count had to be relatively high as well. That Friday, both were low. Since the clinic isn't open on weekends, I had to report to the hospital on Saturday and Sunday to check my blood. On Saturday, my white count was 400 again. Platelets were also still low. More Neupogen shots and another day of waiting followed.

On Sunday, my WBC was 3100/ml and platelets had risen into the acceptable range. The WBC number had gone from 400 to 3100 over night, although it needed to be at least 5000 to start the collection. I figured the number would creep up slowly and I'd see a number just below the required level on Monday which would require another day of waiting before collection could begin. But that kind of thinking was simply anxiety. There would be no guessing involved — I had to wait until Monday to have another measurement.

On Monday, the WBC number had risen into the ball park of 12,000/ml. The number of white cells in my blood had actually gone from lower than normal on Sunday to well above normal on Monday. As soon as that number was recorded, another vial of blood was sent off to the blood bank to assess the number of stem cells present.

This was a separate vial of blood which required a new test specifically to determine if there were sufficient stem cells to begin collection. One of the nurses suggested that she would really be surprised if I were not collecting stem cells the next day (Tuesday). That Tuesday was the day they had tentatively scheduled for me to start stem cell collection. If this last test result produced a good number, they would have planned it perfectly.

The goal of stem cell collection was to collect 10,000,000 (ten million) stem cells. To begin collection, that last special blood test had to count at least 10,000 stem cells per ml of blood.

They sent me back to my hotel that afternoon and said they would call as soon as they received the results. A few hours later, the phone rang and I was told to report to the hospital the next morning for stem cell collection. The analysis of my blood sample showed it contained about 126,000 stem cells/ml. I was worried that it wouldn't reach the 10,000 threshold and it measured more than 126,000/ml. Amazing!

The next morning, I proceeded straight to the hospital to start the collection process. I did some calculations in my head and decided that if my blood had that many stem cells in it, they would only need to separate stem cells from about 100ml of my blood. (126,000/ml × 100ml = 12,600,000)

One hundred milliliters is a relatively small volume of blood. In fact, it is 1/10 of the size of a normal one-liter drip bag of saline IV solution. Maybe the collection day would proceed easy, quickly, and well. Sure, I knew that as stem cells are removed from the blood, the stem cells remaining in the body will be diluted as they evenly redistribute, so the concentration of stem cells in my blood will decrease with time. It is not as simple as just pulling 100ml of blood and separating all stem cells from that blood. (But it was very close to that!)

Each evening after collecting stem cells required another period of "Wait and see!" This time, they sent the whole volume of stem cells collected that day to the blood bank for analysis. They would count the number of stem cells and tell me whether I needed to continue collection the next day. It was a day by day collection process.

Up to five days were planned for stem cell collection. The central venous catheter was a temporary catheter that could remain in place no longer than five days. So they were prepared to continue the collection process for that length of time. Each evening, I went back to the hotel to await the phone call which would tell me what I was doing the next day.

When collection for Day #1 was complete, I went back to my hotel room to await instructions.

Insertion of the Catheter

At this point we need to back up a little to cover the morning of the start of stem cell collection. I arrived at my assigned room at 7 AM. I was told to get comfortable in the bed and the next steps would move along quickly.

Once comfortable in the bed and having had vital signs measured and recorded, they pushed my whole bed out of the room, down the hall, and into the radiology suite. There, I was asked to slide off the bed onto their procedure table.

Then they began to prepare. An oxygen catheter was hung around my ears and nose so I could breathe; they connected an IV drip bag to the catheter in my arm so they could administer medications as needed; and they placed a blood pressure cuff on my right arm which would remain for the duration of the procedure.

The doctor introduced himself and explained the procedure he was about to perform. He needed to make sure I was the right person and he was performing the right procedure. He also questioned me about my medications, medical conditions, family history, etc. When all his questions were answered, he began the procedure.

They covered my whole body with both plastic and cloth sheets. I believe that the only part of me that was exposed to the doctor was a few square inches of skin surrounding the point on my neck where the catheter was to be inserted. The nurse nearest me could see me and talk with me under the upper corner of the sheets. Then, she said she was going to inject something in my IV which would cause me to relax.

I felt a sting at my neck which was the first administration of the local anesthetic they were going to use. I don't remember feeling any pain or any sensations at all in that region of my neck after that one sting.

The next thing I knew, the doctor was asking if I was awake and had I been awake for the whole procedure. It seemed like they couldn't have even started yet, but they were finished. I remember answering that I thought I had been awake throughout, although the nurse suggested that I was snoring at one point, so I probably wasn't awake for the whole procedure.

Then, activity picked up. The sheets that covered me were removed; the oxygen catheter was removed; and I was asked to slide back onto my bed. A few more minutes of being pushed through the halls and I found myself back in my room where I was immediately connected to the stem cell collection device.

There are many holes in my memory of those moments, which means I was groggy and not fully awake during that procedure. What I do remember is that one minute I was covered with a sheet while having the catheter inserted in my neck, and the next minute, I was back in my room having stem cells collected.

Although I had this monstrosity (the outside part of the catheter — everything above the horizontal line in Figure 2, which included all three lumens) attached to and dangling from my neck where I couldn't see it without the aid of a mirror, I don't remember actually feeling it. Long tubes from the two large lumens connected me to the machine. The tubes only required that when I moved around, I had to make sure they were not tangled or crimped in any way. I don't remember any pain or discomfort. It was as if nothing had happened and I was just there relaxing in bed, experiencing an easy, comfortable day.

Although the catheter was anchored to my neck with large stitches, I didn't realize stitches were there until the day when the nurse who was removing the catheter said she needed to cut the stitches to release it. At the end of the collection day, when I was finally allowed to get out of the bed and walk to the bathroom (which had a large mirror above the sink), the nurse had already wrapped the lumens with gauze and taped them to my neck, so the stitches were not visible then either.

All the worries and anxieties of the unknown that preceded the installation of the catheter were for naught. The nurse who kept a constant eye on me during the collection process was pleasant and the whole day proceeded quickly.

When Collection Is Finished, Remove the Catheter!

Believe it or not, with a collection goal of 10 million stem cells and five possible days to complete the process, they collected 11.04 million from me that first day. I was finished in one day, so I could return to the clinic the next morning (Wednesday) to have the catheter removed. Then I could check out of the hotel and go home. Yeah!

I was anxious again, however, because I knew they had used anesthetic to insert the catheter into my neck and I was pretty sure they were **not** going to use anesthetic to remove it. Supposedly these catheters are easy to remove — but that is their word from their experience — and I had never gone through the process myself, so I wasn't sure I could or should believe them.

I know from experience that sticking an IV needle into your arm sometimes goes in painlessly. I also know that when they have removed IV tubes from my arm, sometimes I felt more during removal than I did when they pushed it into my arm in the first place. So I really wasn't sure about the removal of this central venous catheter from my neck. It had to be easy to remove because it could be removed at the clinic by a nurse. Removal didn't require a sterile operating room environment nor a doctor. It had to be easy! Nevertheless, the fear of the unknown caused me to worry.

When I called the clinic the next morning to say I was finished with the stem cell collection process, I was told to come right in. They would see me immediately. Within 5 minutes, I was at the clinic, ready to go. They needed to pull blood samples, change the dressing on the PICC line in my arm, and do a few other things, all of which delayed the removal procedure.

I learned that depending on the blood test results, the nurse had planned to use the larger catheter before removing it. All I needed, however, was some magnesium and they could do that through my PICC line. So the nurse was clear to remove the large catheter.

So here we go! She said that she was going to count to three and on three I should take a deep breath and bear down. I asked what "bear down" meant. I was told, "You know – bear down like you are going to have a bowel movement." That forced me to pause and reflect because the catheter was up under my chin, on my chest, in the front, and nowhere near anything to do with having a bowel movement. I'm sure I looked puzzled and asked several more questions before we proceeded.

I finally decided that she was telling me to tense up and clench my teeth. Do something similar to the grunting fighter pilots must do when they are subjected to high G-forces. Tensing up and grunting helps them through the high stress periods. OK. That made sense. So my mind was still mulling this over and racing at high speed when she asked if I was ready.

She removed the large stitches which held the catheter in place. Then, all that remained was for her to pull it out. Simple! Right?

She said, "Ready?" and started to count to three. On "three" I remember thinking to myself, "Now I'm supposed to take a deep breath." So I did that. Holding my breath wasn't exactly instantaneous with the "three", but I took the deep breath and began to hold it.

I wasn't exactly right on top of things that morning, so everything I did was delayed somewhat from when it was supposed to occur. About the time I was starting to think, "Now ... I'm supposed to tense up and clench my teeth," the nurse interrupted with, "You can breath again. It's out!" I had missed it. I didn't feel a thing. I didn't see her yank on the tube (I had turned my head.) But obviously, on "three" she put pressure on the site on my chest, yanked on the tube, and within a fraction of a second, it was out and she was standing there grinning and waving it in front of me.

No pain. No sensations at all. Actually, her struggles to get the tip of the scissors under the stitches produced more sensations than when she actually yanked the tube out. Once again, I had worried and become anxious for nothing.

Following the removal of the catheter, she needed to maintain pressure on that spot on my neck for several minutes. Maybe ten minutes later, she relaxed the pressure, slowly pulled away, covered the slit and stitch holes with medications and a dressing, taped me up, and said we were finished.

God IS in Control!

I know in my head and in my heart that God is in control. When everyone complimented me on how well "I" had done regarding stem cell mobilization and collection, I knew it wasn't anything I had done, but all credit rightly belonged to God.

If I had control, I would want to gain back the three inches in height that I lost to this cancer. I would concentrate on it, and I'd grow and return to my original height — if I could — but it doesn't work that way. Matthew 6:27 specifically says so. I could concentrate as much as I wanted on making my body cooperate with my meds, and on making my stem cells multiply consistent with the clinic's projections, etc., but I know that all such efforts on my part would have been for naught. I have no control over those things. No human does. God does.

I received all sorts of compliments throughout all of these procedures that really didn't belong to me. I understand that. I think the medical people did, too. God certainly understands.

I know that any cooperation between my body and the nasty combinations of medications I have been taking has been under God's control – not mine. And the fact that there are so many people praying for me and my family throughout this time — certainly some of my successes are due to their intervention to God on my behalf. I had nothing to do with any of this.

The question remains: What are God's plans for me? ... as I go through these chemotherapy treatments? ... and as I move beyond this cancer problem? (... if it is His will that I move beyond this cancer problem.)

I don't know where this is all going to lead. I do know that it appears God has plans for me post-cancer because I have been responding well and continue to respond well to the treatments. If I take everything into account, I have a loving, totally supportive family who is doing their best to see me through this. I have a wonderful, loving wife who is seeing to my every need as she also works to provide me with the best health care insurance in town. I have an expert team of oncologists, transplant doctors, and staff who are guiding me through my treatments. I have a huge support group of Christian friends both locally and far away who are helping in any way possible. Most importantly, they are praying for us and that is wonderful support!

God is in control and He has plans. I don't know what they are yet. I will find out when the time is right. But with the God of the universe in control, I shouldn't have any worries at all. The fact that I worry is due to my human frailties. I know I need not worry — He's got everything under control.

Am I depressed? No, I am not! I am excited and looking forward to getting back to health and to finding out what God has in store. That all brings with it a positive, optimistic attitude.

Even when things don't seem to be going very well, I know God is in control. I am in my 60s. I'm not going to live in this body forever. Some day, changes will cause my body to wind down and I'll become permanently frail. But even then, I know that God will be in control. He has provided wonderfully for me and my family and I have nothing to complain about. He is my activities director and I want Him to lead on. My struggle is to stay out of His way and follow. Lead on, Lord!!

11

High Dose Chemotherapy
& Re-Infusion of Stem Cells

After four cycles of regular chemotherapy, I was scheduled for the "high dose" chemotherapy and stem cell re-infusion in December, 2008. I learned from experience that the physical effects of four chemotherapy cycles are cumulative. Even though the first cycle hardly affected me, each cycle after that was progressively **worse**. By *worse*, I mean that I was weaker and weaker and could do less and less after each cycle. After Cycle #3, I was not looking forward to Cycle #4, which I expected to set new levels of low for **weakness**. And it did! With each successive cycle, I re-evaluated and redefined my definition of what it meant to be **weak**.

I was not particularly worried about the "high dose" week, however, because the oncologist told me that it was not going to be as bad as the normal chemotherapy cycles. So I went into the high dose cycle with the idea that it would be fairly easy.

High Dose Chemotherapy

The "high dose" chemotherapy came and went without incident. I had to move to a hotel about a mile from the clinic which was very convenient. I reported to the transplant section of the clinic each day during that cycle and to the hospital outpatient ward on weekends.

The "high dose" chemo used the drug *melphelan* which (as I understand it) kills every fast reproducing cell in your body. In addition to cancer cells, fast reproducing cells include various blood cells, skin cells, hair follicles, etc.

During the week of "high dose" chemo, they gave me anti-nausea drugs and the melphelan (in that order) by IV on Monday and Tuesday. On Wednesday I received only IV fluids.

I had thought I would be restricted to the suite at the hotel during that first week, but my white blood cell counts were fairly normal. So even though I had received the melphalan, the major effects were yet to come — and I was free to venture from the suite most of that first week.

Stem Cell Re-Infusion

On Thursday of that week, I received my stem cells. They were in two little bags and they didn't take very long to drip through the IV. They gave me fluids again that day and said they needed to keep an eye on me for several hours after the re-infusion. It was another easy day.

Post Infusion Days

That was pretty much IT! Four easy days and the high dose treatment and re-infusion were completed. But the effects were yet to come. Once the high dose medications had been given and the stem cells were re-infused, I was committed to moving forward. At that point, there was no turning back! By that, I mean that there was nothing I could do to change my mind and stop the process. The effects of those four days would proceed whether I liked it or not.

In terms of the numbering of days, the two days of high dose treatments were counted as Days # -3 and # -2. The stem cells were re-infused on Day # 0. So the post stem cell day numbering started from zero on the day of stem cell re-infusion.

Most of the early days post transplant (i.e., post re-infusion) required blood tests and the receiving of IV fluids. They also attached two pumps to the two lumens of my PICC line. The first pump supplied a small amount of anti-nausea medication on a continuous 24/7 basis. It also had a button on it that I could push if I felt nauseous. The pump was set to administer 2ml of the med instantly when the button was pushed. The second pump supplied saline from a 2 liter bag to insure that I was continuously hydrated.

This was good because if I felt nauseous and pushed the button, the 2ml of anti-nausea med would go straight into my blood stream. Normal nauseous feelings occur in your stomach. If you take Phenergan for nausea, you have to swallow some pills which also go into your stomach. Then you must wait for the pills to dissolve to be absorbed into the blood stream. Hopefully, they are not immediately thrown up. But this pump put the medication straight into the blood stream where it was needed. It was great!

I didn't need to push the button but the mere presence of that pump, its medication, and the button were very comforting.

In my case, when I felt nauseous, I needed to stand up, unfold my insides, and belch. A swallow of any kind of soda with its bubbles helped. One day I felt nasty enough that I thought I'd throw up any minute and I could not make myself belch. After taking a swig of a carbonated beverage, however, I stood up and the accompanying belch cleared the pipes. Almost immediately, I felt fine again. It was amazing that my body worked that way, but in my case — belching prevented (and eliminated) any feelings of nausea. This had to do with the way my organs folded when I sat down. They folded funny due to the deterioration of my spine from the MM. So when I stood up, I unfolded and stretched out my insides, which allowed me to burp. If I didn't stand up, pressure built up inside and it felt like nausea.

By Sunday of that first week (which was Day #3), my white cell counts were dropping and at that point, I needed to stay in the suite away from the public. The hotel was within walking distance of the mall, and it was Christmas time — but I wasn't allowed to go out. I was also required to stop eating restaurant food and to eat only things cooked at the suite. We had frozen foods available, so we could cook them for supper and eat sandwiches for breakfast and lunch. Most nights I wasn't very hungry at all. During that time, the mere **thought** of having to stand in the kitchen to cook made me feel sick to my stomach and not very hungry. I never actually became sick from thinking about cooking. On those days, I pushed the cooking duties onto whomever happened to be staying with me at the hotel room.

We ate lasagne, tortellini, turkey pot pies, turkey, beef, and chicken frozen dinners, etc. The food was all quite good, although my appetite wasn't very great.

As the week progressed, my white blood cell count dropped to 100 (Remember: normal is supposed to be 4500 and higher). Also, the number was rounded off, so the WBC count could actually have been as low as 50 and it still would have been rounded up to 100. It was 200 for one day and 100 for the next four days.

During that time, we waited for the white blood cell counts to start growing again. I was as weak during those days as I can ever remember. When lying on my side in bed with a pillow under my arm — I considered it a sign of major weakness that it was difficult to push up onto an elbow to rearrange the pillows and flop down again! Standing up was difficult. Walking was difficult. And during those days, I had a two liter bag of fluid and two pumps in constant tow, so it was like being on a tether. I could only walk so far before the tubing tugged at my shirt and arm to remind me to take along the bag with the pumps and fluids.

The bathroom in the hotel was convenient because it had an outer door and an inner door. The outer room had the sink. The inner room had the toilet and shower. With the outer door closed, one had privacy in the bathroom. The door knob on the inner door was well-positioned to hold the bag containing the pumps and fluids.

Showering was difficult for two reasons. At home the shower has a hose and hand-held spray unit so I could remove it from its overhead position to spray myself. The shower head in the hotel room, however, was fixed — which made it quite difficult to use. Remember: I could not let my upper left arm with the PICC line nor the pump bag get wet. This required holding the bag up over the curtain rod and hanging onto the curtain rod with my left hand during the whole shower. A successful shower required four hands, twisting, and contortions to spray all parts of my body while keeping my left arm and the pump bag up in the air where they could remain dry. It was difficult.

Two nights during the 100 white blood cell (WBC) count period, from midnight to 2AM, I felt rather peppy (relatively speaking of course.) But after going back to sleep, by 5AM when I got up, the peppiness was gone. I then felt as weak as before. On the day my WBC went up to 800, I didn't feel particularly peppy, but I also wasn't tired enough to lay down and take a nap. I was awake and sitting in the living room of the suite for most of that day.

By the next day, which was Day #10, my WBC had risen from 800 to 3900. That crossed the threshold and I was allowed to check out of the hotel and go home.

I still didn't feel particularly peppy, but I was no longer totally weak and unable to do much of anything. I accepted wheelchair rides to and from the transplant rooms on Days #7-9, whereas I had been walking from the front door to the transplant rooms each of the other days. I walked again on Day #10.

Days #5-9 of that cycle were the days during which I was weakest. Those were the days when my WBC, my immune system, and my strength were at their lowest levels. I wouldn't wish that kind of weakness on anyone. At least, in my case, I didn't have to deal with nausea. I did have to deal with diarrhea, though. The high dose chemo was supposed to make you constipated in the days when the melphelan was administered and for a few days following. Then, the effects of the chemo were to go to the other extreme and produce diarrhea.

During the day or two following the stem cell re-infusion, my body was supposed to produce a foul odor (similar to when you have been eating garlic) that would not be noticeable to me, but it would be noticeable to anyone in the vicinity. Everyone that I came in contact with during those days attested to the fact that I smelled funny. This odor is caused by the preservative that is included with the stem cells before they are frozen. As this chemical makes its way through and out of your system, the odor is noticeable. The odor was gone, essentially, by Day #3.

This odor did not affect me at all. I didn't know it was even there and if those around me hadn't commented, I wouldn't have known that I had any odor at all.

Unlike the first four cycles of chemotherapy, the high dose treatment didn't affect my ability to taste. I expected the high dose cycle to change my taste — but it didn't happen. Everything tasted normal. My hunger was affected, though. A little food at meal time was sufficient. My weight still fluctuated wildly due to the fluids they were giving me and to the diarrhea.

The greatest problem I had during this time was weakness. The hotel room near the clinic was a great convenience. It was 5 minutes from the clinic and 20 minutes from the hospital. It would have been a chore to ride 45 minutes each way each day to remain at home during that time. It could be done, but the ride in the car would have been really tiresome.

Once the WBC count was sufficiently high and I was released to live at home again, my energy levels slowly rose towards normal.

Generally, each day following my release from the hotel suite produced a slightly better feeling than the previous day. It was a slow process though, so don't think that the return of energy happens quickly.

12

Recovery

The recovery period follows the re-infusion of stem cells. Remember: the day of the re-infusion is counted as Day #0. With that as the starting point, the recovery period lasts a long time.

Continuation of antibiotic pills lasted beyond Day #360. The stated purpose of those pills was to prevent shingles and a type of pneumonia. Apparently, a person who has already had chicken pox is still immune to it, even after this whole treatment process. But re-exposure to the chicken pox virus can produce shingles. For this reason, they administered anti-viral pills at a reduced level for at least six months.

My transplant nurse told me that I needed to remember that my immune system had taken an extreme jolt from the chemotherapy and stem cell transplant procedures. Because of this, the immune system would be weak and immature for many months following re-infusion. All long-term immunities were zeroed by the process. In computer terms, you could describe the high dose/stem cell transplant procedure as "rebooting" your immune system. This means all of the immunities against common bugs, infections, and viruses that the body had built up over its lifetime, vanished. The body can certainly fight them, but the memory cells in the blood which look for those specific recurring problems are gone. When bugs and viruses are reintroduced, each needs to be fought anew. All memory cells must form again from scratch. That whole process can take years.

As the immune system slowly recovers and matures, immunizations will again be required. These are the same immunizations given to newborns and preschool children. My body may have been 60 but my immune system was zero and starting over.

Around Day #180, I received a vaccination for pneumonia. At the beginning of flu season, I received a flu shot. Later in the fall, I received an H1N1 flu shot. Then, starting a year after the transplant, I was to begin receiving other immunizations.

Weakness

The two major problems with which I had to deal during the recovery phase were (1) weakness and (2) peripheral neuropathy.

In my case, since I did not develop any major fevers (indicating I was free of infection), I went through the whole stem cell transplant procedure as an out-patient. I lived in a hotel near the clinic for the two weeks of high dose chemo and stem cell re-infusion, which was much less expensive than a hospital stay.

During the five days when my white blood cell counts had bottomed out (less than 100), I had little energy. Upon leaving the clinic, I would go back to my room and lay down. That was about as much energy as I could muster. When necessary, I'd get up and walk to the bathroom (about 15 feet) and back.

On Day #11, my white blood cell count was in the 2500/ml range, so I was released from the hotel and allowed to return home. I still had to commute daily to the clinic, but I spent the majority of those days resting at home in Clemson.

On Days #11-13, I felt pretty peppy, relatively speaking. But on Days #14-17, I felt weak again. Blood tests showed that I was *anemic* (too few red blood cells) which had a major effect on my energy levels. On Day #15, I received a transfusion of two units of red blood cells. I felt so much better the next day that I went out shopping with my wife. I had enough energy to ride in the car and be pushed around in the wheel chair, but by the time we returned home, I was whipped. I took a nap as soon as we returned. That helped a little.

On Sunday, Day #17, I didn't feel great again. I don't think I was totally without energy but I didn't feel as peppy as I had felt Monday-Wednesday of that week. My temperature and BP were normal, so it wasn't clear why I felt weak. All I know is that I felt weak again and walking around and even sitting were once again a chore.

The point of all this is to show that energy-wise, I had good days and bad days. The overall weakness that I felt during those days slowly improved with time as I began to exercise and particularly, as I began to walk daily.

Peripheral Neuropathy

My hands and feet felt like pins and needles and were somewhat numb when I started the high dose process. At least three of the chemotherapy medications caused the neuropathy. I was told that the high dose chemical could also adversely affect the neuropathy, which I think it did.

By Days #16-17, my hands and feet hurt over the lengths of my fingers and toes. It wasn't a severe pain — on a scale of 0-10, maybe it was 2-3 — but it was a constant nagging, tingly, numb, feeling of pain. I attributed my walking difficulties to the fact that my feet weren't responding properly to normal signals from my brain to make them move. I fell one day in the hospital because my foot didn't slide properly when my head told it to move to the right. I think my heel lifted and tried to move to the right, but my toes didn't lift and the sole of my shoe dragged and gripped the floor really well. I didn't feel what it did, however. The result was that my foot had not moved at all. So when I shifted my weight to the right, my foot was no longer under me (where I thought it was), and I fell.

The medication I was taking for the neuropathy had been increased to a stronger dose, but I was not able to tell any immediate differences — other than that the neuropathy felt a little worse.

I was told by the oncologist and nurses that neuropathy simply needs time to heal, so at this point in the transplant cycle, it was too soon to tell if anything positive was happening.

Two months after transplant, there was a slight improvement in the neuropathy. By this time, I was walking daily on a tread mill and my strength, ability to walk longer distances, and even my balance had improved. I no longer needed to carry the cane everywhere I went. I was much more stable on my feet just standing around talking with people. But my hands and feet continued to feel numb and tingly.

On Day #60, I told my son that my feet felt especially numb that day. Then a few hours later, I was able to walk and climb stairs more easily than I had been able to do in a long time. Go figure!!

My conclusion agrees with the doctor's words: Only time will tell!!

13

Tandem Stem Cell Transplant

A Second Week of High Dose Chemotherapy
& Another Stem Cell Transplant

At my Day #100 post-transplant visit to see the transplant doctor, I was feeling pretty good. I expected the doctor to look at me and tell me, "You're fine! Go home."

He caught me by surprise when he said, "You're fine! Let's do another cycle of high dose chemotherapy and stem cell transplant."

I knew they had only given me half of the stem cells they collected, but I actually thought they would save those until such time as the cancer had returned. That was an incorrect assumption. According to my primary oncologist, they routinely do a second transplant when the first has gone especially well — as it did in my case.

Apparently, two stem cell transplants, known as "tandem" transplants are standard procedure for multiple myeloma patients receiving the "Arkansas" treatment. I was surprised. I was on my way north to Pennsylvania for the week with my daughter when I learned this. My daughter drove me to the appointment and as soon as I was finished, we headed north. The transplant doctor thought he had already discussed a second transplant with me and he was surprised when I knew nothing about it.

I said I needed to think it over and I'd let him know. I also wanted to talk to my primary oncologist before I decided anything — and that appointment was already scheduled for the following week. I had a week to mull it over.

My biggest objection to a second dose of melphelan and a second stem cell transplant was that my hair, which was finally beginning to return, would disappear again. I didn't like that idea, although I knew it was not a good enough reason to object to the procedure. So I said, "Yes, I'll do it!"

Getting Ready

Legally, before doing such a procedure in South Carolina, they are required to give the patient a battery of tests to insure the patient's health is sufficient to survive another intense procedure. Insurance companies also want prior notice before giving their approval to perform such procedures. The tests and paperwork required several weeks.

I was scheduled to start the new cycle of chemotherapy on a Tuesday, and I was to have another PICC line installed prior to that day. Both had to be postponed for insurance purposes.

Then, I learned that I didn't pass the heart test — the echo-cardiogram. The five previous cycles of chemo had apparently caused my heart function to deteriorate. The value of my left ventricular ejection fraction (its efficiency) was 26%. Normal is 50% and higher. My ejection fraction had decreased over the course of the five cycles of chemo, but I did not know that until this test. One hundred days after transplant, my ejection fraction was too low to proceed. I had the equivalent of congestive heart failure.

To check this test result, they performed a MUGA (Multi-Gated Acquisition) scan. Its results agreed with the echo-cardiogram results.

The whole high dose chemo/stem cell transplant procedure was postponed again and I was sent to a cardiologist.

The Second Transplant

Since I didn't pass the heart test, the second stem cell transplant was postponed indefinitely. I told the transplant doctor that he could keep my stem cells to use when and if "my life depended on it." Until such point in time, I wanted to give the cardiologist and the prescribed heart meds time to work and I wanted to give my heart time to recover.

As I write this (during my second year after the first transplant), my stem cells remain frozen to be used some day (maybe) in the distant future. The transplant doctor continued to monitor my condition. He also set up appointments with my primary oncologist to oversee other post-transplant maintenance procedures.

One year after the transplant, my heart was still recovering. The ejection fraction had returned to 45-50%, which is at the low end of the normal range. Since the ejection fraction returned from 26% to about 50% in two months, the cardiologist wanted to give my heart more time to recover further.

During my second year post-transplant, the second stem cell transplant appears to no longer be a consideration.

14

Heart Problems

Discovery

Since the stem cell transplant, I had been exercising to try to return my body to a more normal state. I borrowed a tread mill and I had been walking 3 miles a day (at a fairly good clip, I may add) without loss of breath. I was not too weak to walk. I had no obvious signs of a problem. My fitness had been continuously improving and I felt much better with each passing post-transplant day.

The "thoracic function test" required in preparation for the second stem cell transplant tested my lung capacity. Those test results showed that my lung capacity had improved since the previous test five months earlier. I was returning to normal.

Then, I flunked the heart function test. Surprise! Surprise!

Left Ventricular Ejection Fraction

I was informed that I had failed the echo-cardiogram — the test that measured my left ventricular ejection fraction. This test measures the efficiency of the final chamber of the heart. It reports the fraction of the blood volume in the left ventrical that is ejected with each heart beat. For example, if my left ventrical can hold 100ml of blood at capacity and it ejects 53ml with each heart beat, that produces an ejection fraction of 0.53 or an efficiency of 53%.

In my case, the measured ejection fraction after five cycles of chemo was 0.26 or 26%. Since normal is considered to be 50% and greater, 26% is too low and the second stem cell transplant was

119

postponed. This number is consistent with the condition known as congestive heart failure. I had acquired this heart problem compliments of my chemotherapy.

At my checkups both before and after my one-year anniversary of the stem cell transplant, my heart's efficiency was still 45-50%. It was still low, but it was at the lower edge of normal — so everyone was pleased.

The meds I was taking for my heart both lowered my BP and my pulse. At about one year plus 25 days post transplant, both my BP and my pulse had risen. In fact, my BP appeared to be accelerating in the upward direction. The cardiologist adjusted dosages and my pulse rate decreased a little, but my BP continued to rise. Then, he suggested he would normally give me a mild diuretic, but wanted me to check with the oncologist before he did so. He was worried about complications in a multiple myeloma patient.

Turns out, I had an appointment with the oncologist the day following the cardiologist's suggested that he'd like to give me a mild diuretic. I read his comment to the oncologist and he agreed. In fact, he said, he would give me a prescription right then and there. I could check with the cardiologist. All were in agreement, so I filled the new prescription.

It strikes me as a little funny that they would give a "water pill" to lower BP, but that was the descriptive name of the new med. The information sheet that accompanied the new med stated that it was primary used to lower BP and that it was indeed a "water pill." Sure enough, my BP lost 10 points the morning after taking the first pill, and it was 10 more points lower the second day. My BP, once again, was back in an OK range. The accelerating BP had stopped and my ankles looked more normal again.

The major swelling in my ankles disappeared after I stopped taking the steroids, but my ankles continued to look puffy even then. The new pills for the BP also helped the puffiness in my ankles. All of these various conditions appear to be related: heart function, BP, pulse, water retention, steroids, chemotherapy, weight gain, etc.

A conclusion is rapidly forming in my mind that the doctors may be chasing after the regulation of all sorts of side-effects in my new post-transplant life. It seems like they get one characteristic under control and another goes haywire.

Several people told me that the body will take months to recover from chemo and especially steroids, and it will take a year or two (at least) for my immune system and regulatory system to regain some sense of normalcy.

One year post-transplant, I was in this stage of my recovery. That is, my body was trying to return to normal after having been bombarded with all sorts of nasty chemicals during the chemo treatments. Some body functions shut down (apparently) in the presence of high doses of cancer meds. It takes time for them to turn back on and reach normal levels after the nasty meds have been removed.

So between one and two years post-transplant, that was my status. Everything was improving slowly. Some functions required medication helps; others did not.

Heart Problems Caused by Chemotherapy

In my studies of this cancer, I don't remember reading, seeing, or hearing anything about possible heart problems resulting from chemotherapy treatments. The possibility of having heart problems following chemotherapy caught me by surprise. I should have known; I should have read somewhere that it was a possibility; but I didn't.

After I failed the echo-cardiogram in March of 2009, I searched specifically for chemo effects on the heart muscle. At that time, I learned that most nasty chemo meds can hurt the heart muscle. The internet pointed to one med in particular which I **had** received 24/7 during chemo weeks. Then, one of the chemo nurses told me that it is probable that all chemo meds can damage the heart muscle.

The point to emphasize here is this: **Yes, chemotherapy meds CAN damage your heart. EVERYONE going through chemo for cancer should be aware of this!**

I wasn't aware! For some reason I didn't know. I missed it! Or maybe the warnings were hidden in the long pages of fine print that listed possible side-effects of each of the various drugs. Only after I was diagnosed with a heart problem did I learn that chemo can damage one's heart. Up to that point, I had not seen any such warnings in the literature.

In any case, I had a heart problem that had to be addressed.

History

The staff had checked my left ventricular ejection fraction three times over the course of my chemotherapy. Normal is 0.50 and greater. My first value was about 0.60, measured in August, 2008 (near the beginning of chemotherapy). Then it was measured again before the stem cell transplant in November, 2008, at which time it was 0.40-0.45. This is a low value, but acceptable for the transplant. In April, 2009, when they were preparing for the second transplant, it was 0.26, which was too low to proceed.

I learned all of these values in April, 2009, after I failed that most recent test. I hadn't seen the numbers before or I might have asked more questions earlier in the process. But that didn't happen. So I was referred to a cardiologist.

Cardiologist

The cardiologist told me that ejection fractions MAY return to acceptable values, but if recovery occurs, it will take a long time. He prescribed two medications to help – one to dilate my blood vessels, and the other to lower my pulse rate. After one adjustment of the dose, my BP dropped to and remained in the vicinity of 130/70, and my resting pulse was around 60. Then, gradually over several months my BP raised to about 140/80 while my pulse remained around 60.

On my third visit to see the cardiologist, I announced that I thought he had hit the dosages quite well — the numbers were pretty good. He agreed. "No changes to dosages. Come back again in two months. And we'll perform another echocardiogram before the next visit."

All this while, both the cardiologist and the transplant doctor looked at me somewhat funny. Apparently, I didn't look like someone with an ejection fraction of 0.26. I didn't feel bad; I wasn't constantly out of breath; my ankles weren't swollen; etc. But the numbers said I had a problem.

Neither doctor told me that I should exercise more because the 3 miles I was walking each day was apparently sufficient. I appeared to be doing the right things, and they encouraged me to continue exercising.

Two months later, during my fourth visit and following another echocardiogram, my ejection fraction had returned to 0.45-0.50 which is at the lower edge of the normal range. This was good news. So he told me to come back again in four months to have another electrocardiogram. No changes to meds were necessary. Whatever I was doing, along with the meds, was fine — continue what I was doing. After the next echocardiogram, my ejection fraction remained at 0.45-0.50. I continued for another 4 months.

Blood Test for the Heart

On my first visit to the cardiologist, he ordered a blood test which would tell how overworked my heart had been. He said he wanted to establish a baseline.

My pulse rate had been 90 and higher for most of the time during and since receiving chemotherapy. I thought it was too high, but there was nothing I could do about it. I thought my heart was being overworked. I was glad to hear that the cardiologist was measuring to see the extent of overwork and damage.

At my second appointment with the cardiologist, he told me the blood test results were normal. A normal result on this blood test suggested that the heart was not damaged and the ejection fraction might return to a more normal value. My heart may have been working hard to circulate sufficient blood throughout my body, but the heart wasn't putting out signs (enzymes) of distress which this test could measure. Great!

Rehabilitation

The rehabilitation phase of my recovery lasted months post-transplant. My first 200 days post-transplant went well. When I started walking on the treadmill at about 50-60 days post-transplant, I could only walk 1/4 mile at a very slow pace (1.5 mph.) Very quickly that was up to 1½ miles at a pace of 4mph twice a day. Then, it became 3 miles at a 4 or 5mph pace once a day.

My strength continued to improve. On occasion, I have walked 4 miles at 4mph, but because I really don't like walking, I usually stop at 3 miles. After walking 4 miles, my legs were tired — but I wasn't out of breath.

Soon after the transplant when we went shopping at a mall, I used the wheelchair. My wife pushed me around. As my walking improved, I started to push my own wheelchair. If anyone asked, I told people it was my walker. When I got tired, I could just sit down. This is still the case — I don't find it embarrassing or unusual to push my own empty wheelchair.

Sometimes, I carry a cane. Soon after the transplant, I needed it for walking. More recently, I use it to maintain my balance and to prop myself up when I'm standing around. On good days, I go without the cane.

When walking around in malls or department stores, I almost always look for comfortable chairs along the way. My wife likes to shop. I like to accompany her, but I'm not necessary to help her shop. So when I don't have the wheelchair, I usually sit in a chair while she shops. Sometimes, I occupy one out in the mall while she shops in the stores.

When she returns, she will find me in the same chair. Sometimes, I have had a nice 30 minute snooze. More often, I drink a latte from a coffee shop while people-watching from a comfortable chair.

I haven't had any problems in public areas. I have been to hospitals, malls, and churches without catching the flu. I have stayed away from schools because I know the kids carry any and all kinds of colds and viruses.

The biggest problem I've had during recovery is a head cold, which dragged on for weeks. It simply wouldn't go away. I usually have one good head cold each winter, but they never lasted as long as the first one post-transplant. I attribute the length of that cold to my lack of immunity.

Recently (400+ days after transplant), I caught a Norovirus which was a 24 hour variety that produced nausea and diarrhea. This virus went through the state and adversely affected many schools. The trickiest part with this virus was that nausea and diarrhea reduce the water in one's body. My water pills also do that. I noticed that my eyes weren't adjusting to the bright sunlight like they should. I immediately measured my BP to realize that it was quite low — it was down to 85/45. I stopped talking the water pills for about 10 days until my BP started to rise beyond normal levels again. Then I resumed the water pills.

That is the problem when you are taking all sorts of pills. Sometimes, natural body functions cause your body to adjust in the same direction the pills were prescribed for. When this happens, it is necessary to catch the fact that too many adjustments are being made in the same direction and to reduce the meds. In this case, the virus and prescription meds both reduced my BP. Fortunately, I am able to measure BP.

I told the oncologist that I had stopped taking the water pills until my BP required I resume taking them. He was pleased I had done so.

Peripheral Neuropathy

The biggest cancer drug side effect I've had to deal with is peripheral neuropathy (PN.) It changed during and after chemotherapy. Sometimes it seemed like it was improving. Sometimes it seemed like it was getting worse. The doctors prescribed gabapentin for the PN. I tried a different PN medication for a while, but that didn't work. About 300 days post-transplant, I stopped taking the PN meds altogether. My hands felt a little better, but my feet and legs felt worse. I stopped the gabapentin because I wanted to know if it was helping or not, and I wanted to stop one of its major side-effects: weight gain.

The main oral cancer drug that I took during the first four cycles of chemotherapy was thalidomide. The maintenance drug (Revlimid®) that I began to take about 150 days post-transplant is a cousin of thalidomide which is supposed to be free of most of thalidomide's side effects. In my case, it appeared to aggravate my neuropathy.

Along with the maintenance drug, I was taking a steroid and a blood thinner. The steroid caused me to be hungry almost all of the time and, as a result, my weight increased. I believe the blood thinner has not cause any problems. It prevented me from having some medical procedures which might have resulted in excessive bleeding, but it did not appear to cause any other problems that I could identify.

About 220 days post-transplant, I told the doctor that the neuropathy in my feet was worsening. It was beginning to move up my legs to my calves, which hurt. I also mentioned that I didn't seem to have any trouble **when taking** the maintenance drug, but my system was royally messed up a few days **after I stopped taking** that drug. The maintenance drugs were to be taken on a 28 day cycle: one pill each day for 21 days, followed by 7 days of no pills. It was during those 7 days that my insides were upset. During those final 7-day periods, I stopped taking the maintenance drugs and the steroids, although I continued to take the blood thinner.

Because my neuropathy was worsening, the doctor told me to stop taking the maintenance meds for the next two months. This also meant I would be stopping the steroids and the blood thinner. Apparently, the upset I experienced during the seven day break wasn't the abrupt stoppage of the maintenance drug, but the abrupt stoppage of the steroid. I learned this by taking some steroids during one of the 7 day rest periods. This time, the doctor prescribed a milder steroid to take during these two months without maintenance drugs.

This new steroid, however, aggravated my water retention problem in my ankles and legs. The steroids appeared to be causing the pain in my feet and legs due to swelling and water retention. Most of these problems went away after I stopped taking steroids.

Water Retention

I began the maintenance drugs right after I had my first visit to the cardiologist when my ejection fraction was identified as being low. My ankles and legs were swelled slightly during that time. I don't remember when I started to notice it. I do remember that both the transplant doctor and the cardiologist noted that my ankles were swollen slightly.

Water retention in ankles and legs can be caused by a weak heart, such as when the heart's ejection fraction is low (which mine was); it can be caused by steroids (which I was taking); and it could come from gabapentin (which I was taking for my peripheral neuropathy.) Since all three were possible causes, it was not clear which one should receive credit for the problem.

In September 2009, I took a break from the cancer maintenance drugs because my neuropathy was bad. In August, just prior to this, the cardiologist had measured my ejection fraction to be 0.45-0.50, which is at the low end of normal. Because my heart was once again functioning in the normal region, my heart should not have been causing the water retention problems in my legs. Since my BP and pulse were good, the cardiologist did not recommend any changes in cardiac meds.

A week later, when I visited the transplant doctor, he prescribed a new, milder steroid. Three days after switching to the new steroid, my ankles and legs had swollen even worse and I developed GI tract problems. At that point, I stopped the steroids completely. About 3 weeks after stopping the steroids, my ankles were nearly normal again. The timing of all of these events indicated to me that the swelling in my ankles and legs was caused by the steroids.

From previous experience, I knew that it took a week or more for the effects of the blood thinner to work their way out of my system. Watching the conditions of my ankles, steroid effects take even longer to work their way out.

About the time of these doctor visits, I had twisted my right ankle slightly. It wasn't sufficient to make me fall, but it was sufficient to make me take notice. When my ankles were swollen, most of the time the two were similar. When they began to swell further with the new steroids, my right ankle and leg were bigger than my left ankle and leg. Both were swollen; the right was worse. I remembered seeing this same symptom the year before when I was in the hospital. Both legs were swollen that weekend, but my right leg was worse than the left. The doctor even ran an ultrasound to look for blood clots in my right leg to identify the cause of the swelling. No clots were found — but I was taking steroids at the time.

Four weeks after stopping all steroids, both of my ankles appeared normal again. I concluded that the water retention problem, in my case, was related to the taking of steroids. I am comfortable that it had nothing to do with the two cardiac meds.

I made a major change in medications at that time. I went from taking the maintenance cancer drug, a steroid, and a blood thinner, to taking none of the three, but a new steroid. After three days of the new steroid, I was taking none of them. That represented a major change.

During the three days I was taking the new steroid, I noticed I was passing blood with my stool. One of my post-transplant standing orders was to pay attention for any unusual bleeding. If I saw some, I was supposed to notify the doctor. So I called.

GI Tract Problems

When he heard that I had bloody stool, the transplant doctor suggested I had hemorrhoids which might have been the cause of the blood. He was about to send me to my family doctor when he asked, "What did your last colonoscopy show?" I had never had one. When he heard that, he decided to set me up with a specialist instead, and he ordered a colonoscopy.

My GI-tract had been in a constant state of turmoil since the time of my cancer diagnosis. When half of the pills I took could cause constipation, the other half could cause diarrhea, and some could cause both, it was no wonder that my insides were upset. But something had triggered a major change, and the trigger appeared to be the change in meds when I stopped taking the cancer maintenance drugs.

By the time I saw the specialist two weeks later, the bleeding was all but gone. He told me that everyone has hemorrhoids so that (plus the blood thinner) may or may not have been the reason for the blood. He also diagnosed a fistula which he said could be fixed after the colonoscopy, if I wanted. I'm still not sure what specific problem caused the blood, but after talking with the specialist, it did **not** appear to be a major problem.

I must admit that I was more anxious about the colonoscopy than about most of the procedures performed on me during the chemotherapy phase of my treatment. Again, it was all wasted anxiety. The day before the procedure, I had to take 32 large pills (20 over an hour at noon and 12 over a half hour at 7PM) which were to clean me out prior to the procedure. The biggest problem that day was not being allowed to eat. I could drink as many clear fluids as I wanted, but I was not allowed to eat anything. It is amazing how great chicken bouillon really tastes!

The day of the colonoscopy, and before the procedure, I wasn't allowed any food or drink. The most difficulty they had in preparation was finding a vein in my arm to place a working IV. As soon as they had the IV taped properly on the back of my hand, I was sedated for 30-40 minutes. I remember wondering how I was going to hold my hands steady for the duration of the procedure, and almost instantly, they were telling me I was finished and asking if I'd like some soda to drink. My wife, who had been out in the waiting room during the procedure, was even sitting next to me already. There was really nothing to it and I was anxious for no good reason.

Now that I have experienced a colonoscopy, I won't be anxious about another one anytime in the future. If you have not had one, I suggest that it is nothing to be worried about.

Cancer-Induced Fatigue

Generally during the rehab period, I felt fine. I didn't have the energy I used to have, however. I cannot really explain it. Apparently, it was a cancer-induced, or cancer-treatment-induced, fatigue. I saw that expression used in a cancer document and it accurately describes the way I felt. I hope it eventually goes away, but I don't know whether it will or not.

I was able to walk several miles without loosing my breath during most of my first year post-transplant; my legs, lungs, and heart must obviously be improving due to the exercise; but I still struggle to find sufficient energy to walk up the steps, or stand up out of a chair. I can do it, but it is not as easy as it once was.

If you must go through a period of chemotherapy, be prepared for weakness and lethargy to follow. I think fatigue goes with the territory for multiple myeloma survivors.

Blood Pressure Control

As mentioned earlier, I stopped taking my high blood pressure meds after the stem cell transplant. I found out 4 months later that my heart had been damaged by the chemo, which caused the low blood pressures.

With the BP under control using medications, I then learned that BP responds to water retention. My bloated ankles and feet returned more or less to normal after quitting the steroids. Some puffiness returned several weeks later. This was accompanied with slowly rising BPs.

I had been monitoring my BP daily for the cardiologist. When the puffiness returned and the BP increased, I took that information to the cardiologist. He suggested both were the result of water retention. I don't know what was causing the water retention, but my climbing BPs seemed to be accelerating — which made me nervous. A mild diuretic was prescribed to solve the BP problem.

The linkage between rising BP and water retention eludes me. According to the oncologist, water retention relates to concentrations of sodium ions in the body. As more fluids are retained, it becomes more difficult for blood to flow easily. Remove the excess fluids; water retention decreases; and BP lowers. I don't totally understand the linkage, but it works.

The diuretic pills came with a detailed explanation sheet. The first line said they were used primarily to control BP. The third sentence said they were "water pills." So water retention and BP are related. The medical community understands this and has pills to solve the problem.

My climbing BP stopped almost immediately upon taking the first diuretic pill. Drops in BP of 10 points each day over the next two days showed me that these two phenomena **are** related.

Chasing Side Effects

This water retention/BP phenomenon suggested to me that the doctors may spend the second year post-transplant chasing secondary effects as my body tries to return to normal. I stopped taking steroids about 9 months post-transplant, and I had stopped a lot of other meds along the way, too. The meds themselves probably flush from one's system within a few days of use. The **effects** of the meds last longer. If the meds happen to overpower the body's normal functions, the body can stop supplying similar chemicals in the presence of high concentrations from pills. It may take the body months to return its own supply functions to normal. For this reason, lots of body systems will continue to act weird long after a stem cell transplant.

As mentioned earlier, all of this depends on one's body constitution and how it responds under the ordeal of chemo. Each individual will respond differently. One simply must wait to see how their body responds, help the doctors to recognize, diagnose, and treat weird behaviors, and hope their body returns quickly to normal.

Remission

The first time I heard the term *complete remission* was about one year post-transplant. Up until then, they may have used the abbreviation "CR" instead. None of the doctors told me straight out, "Your cancer is in complete remission." As I understand the term, however, my cancer has been in complete remission since before Thanksgiving 2008 which was before the stem cell transplant.

After the second stem cell transplant was postponed, I took three cycles of cancer maintenance medications. For multiple myeloma, this was Revlimid®, which is a cousin of the main cancer drug thalidomide. Revlimid® is similar to thalidomide, but it is not supposed to produce all of thalidomide's side effects.

Along with the Revlimid®, one takes a steroid and a blood thinner. The low doses of blood thinner were to prevent deep vein thromboses which were possible when using the maintenance drug.

After about 9 months post-transplant, I was no longer taking any cancer meds or cancer maintenance meds. My one year check up, which included another bone marrow analysis, along with several blood tests, showed no signs of cancer. My primary oncologist was ecstatic. I am ecstatic!

During my second year post-transplant, I continued to recover from the chemo treatments and stem cell transplant.

Re-Immunization

Since my immune system was effectively "rebooted" (to use a computer term) when I was given the high dose medication and stem cell transplant, re-immunization is necessary. Apparently, all immunities built up over my lifetime were cancelled when my body and blood stream were subjected to the high dose of melphelan.

For most of the year following the stem cell transplant, I was taking medications to prevent two major diseases. Apparently, a stem cell transplant patient is not subject again to chicken pox, but new exposure to the chicken pox virus can produce shingles. One of the medications prevents shingles. The second medication prevents a type of pneumonia common to stem cell transplant patients.

Blood tests were scheduled specifically to determine if and when those two medications could be stopped.

The immunization schedule is shown in Table 2. My first post-transplant immunization (around Day 200 post-transplant) was for pneumonia. I received regular and H1N1 flu vaccinations later that fall. Other routine immunizations began after Day 360.

My immunization schedule is shown in Table 2.

Table 2. Immunization Schedule

Post Transplant Vaccination Orders					
Allergies					
The following vaccinations may be obtained at your local Health Department. A listing of local Health Department addresses and telephone numbers are on the back of this form. • You will be responsible for setting up these appointments and any associated fees. • Check with your insurance carrier to see of these vaccines may be a covered benefit. • All patients should be off all immunosuppressive therapy prior to initiation of any vaccination. • Please administer the following vaccines as indicated:					
Vaccine	**200 days post transplant**	**1 year post transplant**	**14 months post transplant**	**2 years post transplant**	**30 months post transplant**
Dates		12/09	02/10	12/10	06/11
Pneumovax (PPV23) 0.5 ml IM	✔			✔	
Tdap (ages 11-64) 0.5 ml IM		✔			
(Td) 0.5 ml IM			✔	✔	
Inactive polio (IPV) 0.5 ml IM		✔	✔	✔	
Haemophilus influenza type b (HIB) 0.5 ml IM		✔	✔	✔	
Measles, Mumps, Rubella (MMR) 0.5 ml SC				✔	✔
Other					

• Patients should receive the influenza vaccine during flu season annually.
• Please administer influenza 0.5ml during seasonal vaccination periods.
• If you have any questions, please call ###-###-####.

Nurse Signature   _Date_

Physician Signature

Section 3

Necessities & Considerations

18

Everyone Needs A Detail Person

It is absolutely necessary that each person facing cancer treatment designates someone to be their "detail" person. In this chapter, I will explain what functions this person performs and the several reasons why someone MUST fill this role.

Definition of the Task

The detail person must keep track of all details of the patient's treatments, prescribed medicines, responses to treatments, medicines taken, and any symptoms and side-effects that present. When the patient is older (e.g., elderly), or younger (e.g., a child), another person must fill this role. As long as the patient is mature and retains their mental acuity, they can perform this task for themselves. Otherwise, a spouse, parent, child, sibling, or close friend could be designated. This is a 24/7 job, so this person must have detailed knowledge of everything that happens relative to the patient, the cancer, and the cancer treatment.

In my case, I have been my own detail person. Were I to lose some of my marbles, I would turn the task over to my wife — but that hasn't happened yet (I think.)

Oral Meds

The Total List of Medications Being Taken

The first absolutely necessary task the detail person MUST perform is to keep track of all details concerning prescription medications.

The orders for my particular chemotherapy treatment program were shown in Table 1. Note that my program combined drugs administered by IV **at the clinic** in combination with prescription oral medications taken **at home**. This was a complex program designed by doctors and administered by chemo nurses and the patient. Administration of oral medications at home should be handled by the detail person. Regardless who distributes the oral meds, the detail person MUST administer and track all relevant details of the patient's pills and other meds. The chemo nurses are responsible for administering IV and oral drugs at the clinic. The detail person is responsible for making sure all medications at home are taken **precisely when and as prescribed**.

The total number of drugs that were used to combat my cancer was a relatively large number — greater than thirty. As detail person for my own treatment, I was responsible for all meds (oral and injected) taken at home. The chemo nurses knew only little about those meds. Some are part of the chemotherapy orders that they follow, so they may check whether or not I had taken a particular pill or had a particular injection at the appropriate time. Some medications were not listed on the chemotherapy orders, so the chemo nurses didn't know about them. Even if they knew all meds prescribed (which they could find in my files), they had no way of knowing which meds **I had actually taken** on any given day — unless they asked. If my mind wasn't all there on any given day, they should be able to get such answers from the detail person.

I saw more than ten different doctors and nurse practitioners during my treatment. Some of their orders and prescriptions were related to the cancer treatment. Some were not. Only the ones related to the cancer treatment were written for the chemo nurses to see. The rest they would have to search for if and when needed. Some details may not have been written at all, so the doctors wouldn't necessarily know all details from my visits to other doctors. Lots of dictations were typed and circulated between doctors, but I don't think all details were always included. The bottom line, in each case, is the detail person. As my own detail person, I knew what each doctor, nurse practitioner, and nurse had said, done, and prescribed, so I could relate those details when asked or when I thought it necessary. The detail person for each patient needs to know all of this so he or she can supply the answers when needed.

During the initial visit to each doctor, during each regular doctor visit, and prior to most procedures performed at the hospital, I had to list all medications (and dosages) I was taking. The detail person needs to have this information available if the patient does not. I carry a copy of a *universal medication form* which lists all of the meds I am taking. It is very useful at times like this.

On occasion, a doctor asked why I was taking a particular medication prescribed by another doctor. The detail person should know the answers to such questions. The answers don't need to be in extreme medical detail or in precise medical terminology. To explain, for example, that "the blood thinner was taken to prevent blood clots in leg veins — a side-effect of one of the cancer meds," was sufficient. Doctors and nurses will understand.

Each patient needs a detail person for medications due to the complexity of chemotherapy regimens, and to the fact that there are very few ways to check whether or not the patient is precisely following the doctors' orders. For this reason, the detail person should be in control of all medications.

Dosage Adjustments

No medication is permanent. When you are supposed to take a particular pill, the quantity, the frequency, and/or the dosage can change from day to day.

Some drugs must be regulated and adjusted according to symptoms or analytical measurements. A side effect of the main cancer drug was the possibility of blood clots forming in the large veins in my thighs. To counter this, they prescribed a blood thinner. The blood thinner, however, needed to be precisely monitored and regulated to produce a very specific blood coagulation time.

I originally took one blood thinner pill each day. During this time, the test recorded an off-scale value — my blood was much too thin. So they reduced my dosage. It was a trial and error adjustment process. They measured my blood's properties and the doctor adjusted the dosage on a weekly basis. I went from a whole pill a day to half a pill a day, to half a pill every other day, etc. Finally, we arrived at a level that produced the desired results.

During this adjustment process, the doctor used the measurements and input from the detail person (me) to help decide if and how to adjust the dosage.

"As Needed" Medications

Some drugs are to be taken only *as needed*. I carry several bottles of pills that are in this category. This includes pain pills, stool softeners, laxatives, pills that combat diarrhea and nausea, etc. I even had some insulin in the refrigerator in case I needed it.

My neuropathy pills were to be taken *as needed*. I learned that those particular pills were not quick response pills — that is, for best effect, they had to be taken for several weeks before their effects could be felt. Waiting until my hands and feet hurt before taking them probably would have produced little, if any, improvement. For that reason, I decided that I needed to start taking them regularly. Several weeks later, the nurse suggested that I was taking a particularly weak dosage — so it was increased. Three months later, because of side-effects, I began to reduce the dose and wean myself off those pills. The detail person needs to know all of this.

There are also mineral supplements (magnesium and potassium, for example) that were to be administered in response to measured blood chemistries. They prescribed potassium and magnesium pills for me when both magnesium and potassium levels in my blood were low following chemotherapy. For a while, every time I visited the clinic, they measured my blood chemistry to determine if I needed magnesium and potassium ions along with the liter bag of IV saline. When I needed the extra ions, I received them. They were administered *as needed*.

The dosages of those ions in IV form were adjusted in response to my blood chemistries. This was supplemented with magnesium and potassium pills. Since the doctors and nurses carefully monitored these ions in my blood, the dosages changed frequently.

Pain killers were also prescribed to be taken *as needed*. In most cases, pain killers do respond quickly, so when various body parts ache or when sharp pains occur, I can take some pain killers and expect relief within the hour.

In all of these cases, the detail person needs to know which pills have been prescribed, which have been taken, when they were taken, and how many were taken. He or she must be able to answer any and all questions regarding meds (1) prescribed and (2) actually taken.

Doctor Visits

The detail person also needs to accompany the patient to all doctor visits. The first reason, of course, is to insure on-time arrivals for all appointments.

The nurses and doctors will always ask questions like, "How are you doing this week?" or "Do you have any pain today?" or "On a scale of 0 to 10, how much pain are you experiencing?" The patient's answers and the detail person's answers to such questions may differ markedly. Both will be of use to the doctor.

During and after Cycle #4 of chemo, I felt generally awful! When asked how I was doing, I answered, "I feel terrible! On an absolute scale, I am probably doing perfectly fine." The doctor had to sort that one out for himself. I think he understood. Had my wife been present, she could have given him her opinions of how I was doing — and they probably would have differed from mine.

Throughout my chemotherapy treatments, the doctors and nurses all constantly told me how well I was doing. One day when I felt particularly terrible, the nurse practitioner at the cancer clinic told me, "You look great!" In response, I told her that I didn't feel great, and when I looked in a mirror, I didn't look great either. But if I compared myself to others I saw in the waiting room, I felt fine.

You, the cancer patient, will answer all questions relative to yourself. The doctors and nurses will evaluate your condition relative to all of the other patients who walk through their doors. I am sure, from their responses during most of my chemotherapy sessions, that I looked like little if anything had happened to me (relatively speaking, of course). I am also sure they routinely saw people who were in much worse condition than I was. So I felt blessed — even when I felt awful — because I understood how others responded to these same treatments.

My wife may have a better handle on my level of fatigue, my lack of energy, and my overall state of well-being than I do. She notices quickly when my balance is off, when my hands shake, or when I can't stand still or walk without falling. She knows when I am sleeping well (or not at all), how I am eating, and generally how everything else is going. Her observations are every bit as useful to the doctor as mine. When I feel well, I drive myself to the clinic. On days when I don't feel well, my wife takes me — which means she is present to answer questions from her point-of-view.

The detail person also needs to keep track of what doctors do, say, and order. For this reason, the detail person should be present to hear all of the doctors' comments. Keeping notes of doctors' comments is advisable as well. If a doctor says during a visit, "Next time, we need to readjust the dosage of a particular medication," the detail person should make note of it. If necessary during the next visit, the detail person can then remind the doctor of that earlier comment.

After one procedure, the nurse told my wife that I would be particularly forgetful and maybe even confused for the rest of the day. She should therefore keep an eye on me and take all of my comments in the proper light. I insisted that I was not overly forgetful that day, but my wife disagreed.

I am naturally a detail person. If I were no longer in my right mind, or if my memory had been affected by my cancer, I understand that someone else – my wife, daughter, son, or a friend — would **need** to be the detail person. On any given day, several people can share in the duties of detail person, but **ONE** person ultimately **MUST** compile and keep track of all information and be responsible.

Do not assume that your doctor and his nurses will function as your detail person. Neither the doctors nor the nurses are with you 24/7 so it is not possible that they can know all appropriate details. **Most** of your information will be available in their records (this applies especially to hospital inpatients), so when a doctor examines you, he/she can scan the records and learn more details than you could imagine. But such detailed record keeping doesn't happen for patients living at home. There is no possible record that will tell a doctor if you took this morning's required meds. ... or last night's. ... or if you are in severe pain, etc. In the hospital, nurses write down every little detail. Outside the hospital, nothing is recorded — so the detail person needs to know this information.

You need to decide whether you can do this job all by yourself, and if not, you need to designate someone to the task. You may need help making this decision. I can think of many people who would choose to be their own detail person — even when they are **not** capable of handling the task. This applies especially to older people who are beginning to show signs of dementia. They think they are fully capable of taking care of themselves, when they are not. In such cases, a responsible family member or friend must step in and make the decision.

The designee must be a responsible person who spends lots of time with the patient. A spouse, parent, adult child, or a close friend can perform this function. But someone **MUST** do it! It is not a position that can be ignored.

What is the expected result? ... between the patient, the detail person, the doctor, and the whole medical team, chances for a successful outcome of the chosen treatment will be maximized.

Overall Details

The detail person needs to know both what you are **supposed** to be doing and what you **are** doing. As I have said, I am the detail person for my battle with cancer. As an engineer, computer programmer, professor, and teacher, I have always been a detail person. I think that way. I want to know details about anything and everything. So I am the logical person to play this role for my own battle with cancer.

Getting Organized

Because I am a very organized person, early in this process, I purchased a pill container that holds a week's worth of pills. Each day has compartments labeled 'breakfast,' 'lunch', 'dinner,' and 'bedtime.' Seven groups of compartments (one for each day of the week) make up the whole container. Every Sunday evening or Monday morning, I sit down with the empty container and with my bag of pills to dispense all pills to be taken during the following week.

I know what each pill does; I know why I am taking each med; and I can generally identify each pill by its shape and color. No one else in the family can do this because no one else in the family has anything to do with my pills. On a good day, some of the compartments in my pill container are empty. Other times, some of the compartments have 14 or more pills in them. I take some pills every day, some every other day, some Monday, Wednesday, and Friday only, some Saturday and Sunday only, some during the four days of the chemotherapy cycle only, etc.

Had I designated someone else to do this for me, I would still be paying close attention to the details I just mentioned. If I were not able to fulfill this task, I would hand it off to someone else and trust that person to do everything accurately. The detail person needs to be accurate and precise in their dispensing of all meds. The detail person must be someone you trust implicitly! It is a critically important job!

Daily Assessments

In addition to one's meds, other assessments must be made on a regular basis. Some assessments can be made using simple tools like a thermometer and a blood pressure cuff.

It is somewhat funny, I think, that for sixty years I never owned a blood pressure cuff. I was an EMT when we lived in the upstate of New York, and I borrowed a blood pressure cuff to use during my whole time of active participation. In response to the oncologist's comments, I wanted to be able to measure my own BP. One Sunday, in late 2008, I went to the local drug store and purchased a BP cuff. After sixty years, I finally owned one.

The very next day, when I went to the transplant clinic, the nurse handed me a box. She said that they always give each of their new transplant patients this gift at the start of their program. The box contained a BP cuff. I never owned one for all those years — and when I finally decided I needed one, and purchased one — the very next day they gave me one. I immediately returned mine to the local drug store for a full refund, so today I own one — only one.

With the results from the two instruments (thermometer and BP cuff) and a subjective assessment of how the patient feels from day to day, the detail person can decide when and if there is a problem and when and if an emergency call should be placed to the doctor.

The cancer clinic staff gave me a list of problems, which, if they exist, should prompt me to call the doctor. A temperature greater than

100.5° indicates an infection. Call the doctor. BP is too high or too low. Call. You can't walk or stand up without falling. Call. Unexplained bleeding. Call. Nausea or diarrhea. Call. Etc.

As my own detail person, I provide input on each of these to other family members, but when a problem actually exists and a call needs to be made, I do it if I am able. If not, my wife places the call. Even though I do not consider my wife to be my official detail person, she can take one look at me, evaluate how I am doing, make her own assessment, and recommend that I should call or place the call herself.

Sometimes, one of the problems on the list exists and the patient doesn't realize it. My son walked over to me one day and asked why the white of my eye was actually red? I didn't realize it. Looking in a mirror, sure enough, the white of my left eye was bright red. It was unexplained bleeding — which is on the list. So I called.

On the days when my BP was extremely low and I could not walk, my wife made the decision for me. She called the doctor. Actually, the day she called, I also had a slight fever, so she had two reasons to call. They admitted me to the hospital that afternoon. They were rather speedy about it. I didn't have time to pass GO or collect $200. I managed to climb into the car and ride to the clinic where they took one look at me and told me, "We're admitting you to the hospital. Your wife can drive you there." Within five minutes, I had a room number and I was told that the hospital staff were expecting me. My wife simply had to make that first decision to call the clinic and everything moved quickly from there.

I remember not feeling well that day. Walking was difficult, BP was low, and I had a fever. The last thing I wanted to do was think and make decisions. My wife made those decisions and took care of everything. On that day, she functioned as the detail person and the decision maker and I was happy for her to do so.

Summary

The main point is this: **Every patient needs a detail person who is keeping track of everything.**

This detail person needs to keep track of:
- everything the doctors say,

- everything the nurses say,
- everything the patient says,
- all details concerning how the patient feels,
- all medications the patient **should be taking**, and
- all medications the patient **has actually taken**.

If at all possible, the detail person should accompany the patient to all office visits, chemo sessions, and to all other testing procedures.

The detail person is **an absolute necessity for everyone who has cancer (or any other serious illness.)**

19

Care Giver(s)

One requirement for the stem cell transplant procedure is to have a person designated as the **Care Giver**. In my case, my permanent primary Care Giver is my wife. But the requirements for this job sometimes make it difficult for one person to be the designated Care Giver 24/7.

During chemotherapy treatments and especially during high dose chemo and stem cell transplant procedures, the Care Giver needs to be available 24/7.

The Designated Care Giver

A single person needs to be designated as the patient's Care Giver for the whole cancer treatment process. This doesn't mean, however, that the person must quit his or her job. This person will most likely be the person to notify in case of an emergency. This person may function as the Detail Person discussed in the previous chapter. But other people can and should help in the capacity as Care Giver.

Depending upon the patient's level of capability or incapacitation during treatment procedures, the Care Giver may only need **to be present** if and when needed. This was the case with some of my treatments. My wife was the responsible person in all cases, but she continued working her job. Evenings at home, on her days off, and when her schedule permitted, she was physically present with me, fulfilling the role of Care Giver. The rest of the time, family and friends helped fill this role.

Many procedures and office visits required a driver to take me to the appointment. My wife, children, and friends performed these tasks.

On days when we knew I would be at a procedure for several hours, sometimes one person drove me to the appointment and another picked me up later in the day. When I was at home and my wife was at work, one of my sons was present with me.

When I had to move to a hotel room in Greenville for the stem cell transplant procedures, my wife and sons took turns as Care Giver for the first stay, and my wife and a nephew took turns as Care Giver for the second stay. I was **not** incapacitated that I needed constant help, assistance, or attention during any of the procedures, but someone had to be present in case I did become incapacitated, or I did need help, or I did need an immediate ride to the clinic or hospital. My nephew, who was present through most of my high dose chemo/stem cell transplant procedure was generally bored — but I couldn't have done it without his presence.

In each of the two transplant procedures (harvesting stem cells, and administering the high dose chemo and reinfusion of stem cells) when I had to stay in Greenville, I had a two-bedroom suite at a hotel near the cancer clinic. The Care Giver stayed in one room while I stayed in the other. The Care Giver sometimes prepared meals, but mostly they were present and available only to help as needed 24/7. Each morning, they drove me to the clinic or to the hospital and then they were free until pickup time. Sometimes they dropped me off, drove home to Clemson, and a different person picked me up later that day. Sometimes they simply returned to the suite until I called to say I was finished. This allowed my wife to keep her job throughout the whole treatment process.

Sometimes, I needed a ride from Clemson to Seneca (10 miles) or to Greenville (35 miles) in the morning and a return ride in the afternoon. Sometimes I had multiple appointments scheduled in both Seneca and Greenville on the same day. My family and friends always made sure I arrived on time for all of my appointments. In each case, the person who accompanied me was fulfilling the task of Care Giver at that particular point in time. If I needed any type of help, they were there to assist.

Each person who goes through the cancer treatment process needs to look into Care Giver requirements by their particular doctors, clinic, hospital, and health insurer. Requirements can differ, so you need to ask questions to find out the local rules, regulations, and requirements.

It was simply not allowed that I spend 18 of every 24 hours in my hotel room **by myself**. Someone was **required** to be with me — a Care Giver — even if they did nothing but watch TV and sleep.

To Wait Or Not To Wait

At each appointment at a clinic, hospital, or doctor's office, the patient should always ask if the Care Giver needs to wait around, or if they are free to leave for the duration of the appointment. It would be an enormous waste of time for a Care Giver to sit in the waiting room when the patient is under the care of nurses and doctors. Ask! Don't automatically assume that someone must wait around.

For example, chemotherapy at the clinic usually required several hours. If it was an hour drive from home to the clinic, the driver usually made the decision whether to wait around or to leave and return later. Stem cell collection procedures were described as one or more all-day (8 hour) procedures at the hospital. Blood transfusion days were long days. Chemotherapy days with accompanying office visits were generally long. Most high dose chemo and post-transplant days were long. Some analysis procedures, such as MRI, were long. In each such case, we tried to determine in advance how long the visit would be so the driver could decide how he or she would spend their day.

In many cases, the driver was free to go shopping or otherwise occupy their time elsewhere until the designated pickup time. Cell phones helped ease any problems. When the nurse said I would be free to leave in about 15 minutes, I could call my Care Giver to notify them of my expected departure time. On days when my nephew was waiting at our hotel room about 5 minutes away from the clinic, I could call him when I was about to be released. By the time I walked to the front door, he was usually pulling into the parking lot.

Plan Ahead

As much as possible, plan ahead. Keep track of family and friends who are available on given days. Try to spread the driving around — especially with friends. Don't overwork any one friend if at all possible.

I learned that one friend was disappointed that she wasn't needed to drive me to Greenville for two appointments on the day I had the fever and was admitted to the hospital. After hearing that, I made sure to ask her if she could drive me to Greenville for the rescheduled appointments.

One might think that such needs are a pain for everyone involved, but I learned that when friends tell you they want to help — it generally means they **want** to help! I had more friends volunteer to help than I actually had needs available for them to do. And like this example, some can be disappointed when they don't have the opportunity to help.

Everything worked out well for me throughout the whole process. I was never home alone (other than for a few minutes here or there.) Even more recently when I am feeling much better and it appears that I am alone, almost always one of the boys is around the house somewhere (as college students, they tend to keep strange hours.)

I have been blessed with excellent Care Givers throughout my treatment process. I have been especially fortunate on some days when I really needed a Care Giver — that my wife was able to be with me.

20

Emotions

A person's emotions and their hunger, weight, and diet (next chapter) can swing over the whole spectrum of possibilities. I do not know precisely how everyone else will react on these points because everyone is different. I can only describe my own reactions.

In my case, my emotions swung from one end of the spectrum to the other during my chemotherapy treatments. For my first 60 years, I would characterize myself as emotional but with the ability to keep it inside and under control.

Throughout most of my life, I hardly ever had to deal with my emotions running wild. If I watched a tear-jerker movie, I might have tears in my eyes at the most intense parts, but that's it. Now that I have had cancer and have gone through intense chemotherapy, I tear up easily at the least intense happy, sad, or emotional moments.

When friends tell me they are praying for me, or when they say they want to pray for me, or when the kids from school send cards that say "We're praying for you", I tear up — and at such moments, I can't talk. When gifts arrive from Christian friends, I tear up. I am not accustomed to receiving gifts, offers of help, or prayers on my behalf like have occurred frequently and constantly throughout my illness. I do not deserve such gestures of grace and love, so it is very emotional when they arrive. I know that God is taking care of my family and me throughout this battle but I am still surprised when He does it and when He shows His abundant grace through His people.

The unsolicited grace of God is somewhat unusual to me because I have been accustomed to having a biweekly paycheck arrive from my employment. I have always felt that my places of employment were the results of God's direction in our lives, but the fact that we could count on a biweekly check that was more than sufficient to meet our needs meant that we were one step removed from relying on God's gracious provisions on a daily basis.

Going from a university professorship to a consulting job in the Far East to less consulting here in the US to teaching middle and high school science at a private Christian school to being totally disabled each represented successive drops in income that helped prepare us for this time of disability. My income didn't drop from high to low to non-existent in one giant step — it was a somewhat gradual decline — which caused us to rely more and more heavily on the Lord's grace over a period of time.

I should have been thinking this way all of my life, but I didn't. Now that I'm in my 60s, I'm learning something I should have learned much earlier in my Christian life. Later, however, is better than never!

It is still very emotional to receive gifts that God has provided. We never know when and from where they might be coming. But He knows His options and He has it all under control. And it brings tears to my eyes when it happens.

My emotions are right there — out in front now. I don't know if they'll recede again as I move further and further from the intense days of chemotherapy. That remains to be seen. But for the time being, they are rather intense and frequently visible for all to see.

I assume this is completely normal for cancer/chemotherapy/stem cell transplant patients. I do not know which of these three specifically is the cause of the change — but since I've gone through all three, I attribute the emotional changes to the whole process.

21

Hunger, Weight, & Diet

During the times of treatment of this cancer, I have experienced a wide range of variations in my eating, hunger, diet patterns, and body weight. I will break down the time periods into five parts: **before** diagnosis, during the **first** two cycles of chemotherapy, during the **second** two cycles, during **high dose** treatments, and **after** completion of all chemotherapy.

Before Diagnosis

While my body was beginning to deal with multiple myeloma, but before I knew what was happening to me, I can define two time periods: when I had no pain, and when the pain was developing.

Before Any Pains Developed

During the fall of 2007, I was dieting and had gotten my weight down to my target weight by Christmas time. Things were going well and I was without any pains to indicate a problem.

I had experienced what appeared to be a cracked rib but that pain went away within a few weeks. Throughout most of that fall, there was no pain and the diet proceeded successfully. At Christmas, I was near the weight my family doctor told me I should achieve and hold.

As Pains Were Developing

After Christmas, pains in my rib area and abdomen started to return. Although I could not recall having done anything to aggravate my injured rib, the pains came back.

As I became more and more uncomfortable, I lost no more weight. I started sitting around more and moving less. At school, I identified the most comfortable chairs and sat in them more and more often.

Movement brought pain. Sitting still in a comfortable chair eliminated the pain. Getting into a chair was fairly simple. Getting out of chairs was much more difficult. But sitting motionless was accompanied by painless comfort. So I spent more and more time sitting to minimize pain and maximize comfort.

I must admit that when I feel poorly or when I hurt, I eat. My rationale says that if I'm going to hurt, I am at least going to be comfortably full. So I began to eat anything and everything that sounded good. Frequently I had milkshakes on the way home from school. I ate cheeseburgers and french fries, candy, ice cream, and anything else that satisfied my cravings.

By the time I was diagnosed with cancer, I had gained back about twenty of the pounds I lost the previous fall.

During First Two Chemo Cycles

The fact that the cancer medications that accompanied chemotherapy treatments were very complex meant that my weight could go either way — up or down. It all depended how my body would respond.

I knew that if I became constantly nauseous, I could lose weight. If I developed diarrhea, I could lose weight. If my hunger kicked in, I could gain weight.

Fortunately, I only had nausea and diarrhea problems during one day of the first cycle of chemo. I learned quickly that I had no energy whatsoever that day. It lasted only one day and I credit having eaten too much the night before for that awful day. After that, I was more careful about what and how much I ate.

I also learned from my experiences with MM and with the deterioration of my vertebrae that sitting folded my insides in a somewhat unusual way. I suggested this to two of the doctors and they agreed that my insides probably **were** folding in an unusual way — although they said it was normal for MM patients. During that time, I learned that if I could belch, I could minimize any feelings of nausea. I learned that if I could sit up straight or stand up straight and unfold my insides, which usually allowed me to belch, nauseous feelings went away. So if I felt the least bit questionable about nausea, I would stand up to unfold my insides and try to belch. I might also drink some water or soda to force a belch. That usually solved any problems.

The problem with the complexity of medications I was taking was that there were several meds which had possible side effects in opposition to one another. Some could cause diarrhea — others constipation. Some could cause loss of hunger — others revved up one's hunger. Etc. In some cases, because a particular med could cause certain bad side effects, they gave me a second med to counteract that first one's bad side effects. The cancer meds, for example, could cause blood clots in veins. So I took blood thinners as a precaution.

What happens during such times is a complex interaction between one's constitution, the effects of the cancer, the complexity of the large number of drugs being taken, and the way the body responds to everything.

In my case, the first cycle of chemotherapy hardly affected me. I wasn't particularly hungry during most of that cycle and I lost weight.

During the second cycle, I continued to have low appetite and I continued to lose weight. Then, during the second week of the second cycle, my BP was low, and I developed an infection.

My blood pressure was lowered both by the chemo treatments and by my blood pressure meds (to combat high blood pressure). The combination of the two lowered my BP too far — to the point that I became light-headed and weak, and I could not walk without help.

When my body temperature went up over 100°F, it indicated that my body was also fighting an infection. This was the day I was hospitalized: the Thursday before Labor Day weekend of 2008. I learned the next morning that I would be spending the whole weekend in the hospital.

In the hospital, they provided a food tray at each meal — so the volume I was given was limited. My weight continued to decrease to the point that I was back in my target range where I had been the previous Christmas. I thought maybe my weight losses would continue beyond that point. Some of my friends thought I would lose weight precipitously. But I did not.

During Second Two Chemo Cycles

The steroids that I was taking in relatively concentrated form during chemo weeks were revving up my appetite. I was told that they would put lots of sugar into my blood stream, so I was also taking another med to help remove excess sugar from my blood stream.

During the third and fourth cycles of chemotherapy, my appetite came back in force. I was hungry, even when I knew I was full. I gained about ten pounds during the third and beginning of the fourth cycles. Then, the steroids of the fourth cycle caused my body to hold water. I gained about ten pounds of water weight in 48 hours. I lost some of the water and its weight in the following days. Eventually I lost most of the excess water (the bloating disappeared), but the majority of the weight remained.

Because my legs swelled up with water, they gave me diuretics to remove the water. The diuretics didn't work well. I lost water quickly within several hours of each dose of diuretic, but my overall weight stayed up for some reason, and my legs remained swollen.

After the fourth chemo cycle, it was November 2008, and I was given some time to relax before the fifth, high dose cycle. By then, my body had regained the twenty pounds I had lost twice already.

During High Dose Chemo Treatment

The high dose treatments were not too bad. The transplant staff at the clinic gave me lots of fluids along with the meds, so my weight typically went up during the morning treatments and back down later in the day.

But they warned me that the melphelan (the high dose medication) tended to make one constipated initially and then, longer term, it tended to give one diarrhea. Both extremes were once again possible. If constipation was to happen, it would occur almost immediately. Diarrhea could occur later — at the end of the first week and during the second and third weeks following treatment.

I didn't experience constipation, but I had to deal with mild diarrhea for several days. Even so, my weight remained high following the high dose treatment and stem cell transplant.

After the Completion of All Chemotherapy

At this point in the process, when my body was attempting to return to 'normal', hunger and weight returned to normal as well. I have always been overweight — I was still overweight. I didn't lose lots of weight during chemotherapy like I thought I would. In fact, I had gained. So after the completion of chemotherapy, my goals were to exercise more, build up muscles that had grown weak, and lose weight.

Neuropathy Meds

By 120 days post-transplant, my weight was up about 40 pounds from my target weight which I had achieved twice in the previous two years. I mentioned to a friend that my weight was going the wrong way and I didn't like it. He commented that I was taking meds for neuropathy — and they could cause me to gain weight.

Sure enough, I read the neuropathy medication side-effects list carefully and *weight gain* was on it. My friend suggested that when his wife went off the neuropathy meds, she lost 40 pounds (and she is tiny.) That just happened to be the same number I was working with — so there was hope that when I was no longer taking that medication, my weight might again decrease towards normal.

Steroids

A lot of the weight gains have been the result of the steroids I have taken since the start of chemotherapy. The main steroid taken, prednisone, was prescribed originally as a pain killer. Other steroids were prescribed in concentrated doses during chemo weeks, and the prednisone was prescribed to be taken with the cancer maintenance meds.

All told, I took steroids regularly for about 15 months since the start of chemo treatments. I stopped all steroids about 9 months post-transplant because my legs were swelling with excess water. A milder form of steroid was prescribed to replace the prednisone, but that increased the bloating in my legs. At that point, I weaned myself off of all steroids (over about a week's time). By my one-year checkup post-transplant, I had been off the steroids for about 3 months.

I still was having mild bloating problems in my legs and ankles, but the bigger problem was when my BP started to rise. I noticed that my ankles were still slightly bloated, I was going to the bathroom every two hours or so (day and night), and my BP was rising on a daily basis. My meds and activities had remained unchanged for several weeks, so when the only change I could detect was rising BP — I began to worry. The cardiologist changed a dosage and told me that should help — but if it didn't, I should let him know. When I e-mailed him that the BP was continuing to rise, he suggested a diuretic, but also suggested I check it with my oncologist. They both agreed.

So I filled the prescription for a diuretic, which the paperwork described as a "water pill". The oncologist said that my body was holding too many sodium ions which in turn were holding water. The diuretic would remove some of those sodium ions along with the attached water, and my BP should decrease. Sure enough, my BP dropped 10 points each of the first two days on the new meds, and it continued to decrease slowly after that. The bloating has decreased and I can sleep through the whole night again. My weight is still high, but the excess water seems to be under control.

I read the list of side effects for the "water pills," and practically everything is listed. Those tiny pills can really mess up the body until it adjusts to the new med. They threw my insides out of whack again, but with time, the crazy behavior began to subside towards normal.

Summary

Hunger, weight, and diet changes occurred throughout cancer treatments. Body weight can go up or down, depending on the particular meds taken and your particular body constitution. The result is unpredictable.

Some of my friends thought they would no longer recognize me after my chemotherapy because I would lose tons of weight — but this did not happen.

My experience suggests that everyone should be prepared for a roller coaster ride in these three areas!

Other Observations & Pointers

During the course of my treatment, I observed how my body reacted to chemotherapy and various other procedures to which I was subjected. Many of these observations and considerations may be useful to someone who needs to go through chemotherapy (recognizing that all individual treatments will differ), so they are written here for your information.

Disclaimer

Note that I am NOT a medical doctor. My PhD is in Solid State Science (materials science). Over the years, I have studied biology and chemistry (mostly inorganic chemistry); as an Assistant Fire Chief with responsibilities over the ambulance squad in the small village of Alfred, New York, I was an emergency medical technician (EMT); and more recently I taught biology at the middle and high school levels. Little did I realize during any of those times that the experiences would help prepare me to deal with this cancer.

Since I am NOT a medical doctor, any suggestions that follow come from a knowledgeable observer and fellow cancer patient — NOT from a medical expert. If any of these suggestions sounds like a good idea, discuss it with the appropriate physician(s) or nurses. DO NOT simply take my word for anything.

Everything that follows is the result of my observations, my experiences, my biases, etc., from my dealings with my cancer (multiple myeloma.) I am NOT making recommendations as an MD. I am noting my experiences and trying to remember what worked and what didn't work for me as I went through this ordeal. Notice all the 'me's, 'my's, and 'I's in the preceding sentences. Cancer treatment and chemotherapy is an individual experience. What worked for me may or may not work for another. The knowledge of my experiences may be extremely valuable information for others to know in preparation for their own chemotherapy and their own battle with cancer. For these reasons, I offer my observations, considerations, and pointers in this chapter.

Also, it would have been a big help if I had known in advance the kinds of procedures I would be subjected to as I went through chemo treatments. I didn't know much of anything and I learned everything as I experienced it. The topics I have included in this book would have been a great help to me had I been able to read them in advance. Hopefully, my experiences can now be a help to others.

Read! Read! Read!

I recommend that everyone who has to deal with a particular cancer reads and learns as much as they can about that cancer. The sooner after diagnosis this is done, the better!

I mentioned a few chapters ago that everyone needs a detail person for their treatment process. Whomever is designated to be detail person, it behooves both that person and the patient to **read, read, read**. Each cancer is different and each can affect different people in different ways. Each cancer will have its own treatment, and each treatment will produce direct effects as well as a variety of side-effects. The more well-read both the patient and detail person, the more sense can be made out of the quirks of each particular cancer and the details of its treatment process.

Comfortable Place to Spend Time

I recommend that each cancer sufferer identify a comfortable place to spend their time. In my case, early in the treatment process, we purchased a recliner for our sun room. This chair was sufficiently comfortable for both sitting and sleeping. All of my pills and notebooks, the thermometer and blood pressure cuff, correspondence materials, reading materials, files, checkbook, calculator, etc., were located within arm's reach of the recliner. This worked well! It became my corner of the house.

I had never before owned a recliner, but it is one of the most comfortable chairs in the whole house and one of the best investments I have ever made. As I suggested, this chair also frequently served as an alternate place to sleep.

Finally, make sure that your comfortable place is located in an area frequented by other family members. The last thing you want is to have a comfortable seat off in a corner where no one ever visits.

Sleeping Arrangements

This may sound weird because everyone usually has a bed in which to sleep, but it is helpful to identify other possible sleeping locations throughout the house. For a while at the beginning of treatments, I could not sleep in bed on my stomach, which is my normal position, nor was it possible to roll over and sleep on my back. I simply could not lay down in any position. The damage to my spine receives credit for this.

During that time, I spent many nights sleeping in the recliner. Since it is not completely flat, I could sleep on my back with my knees and waist slightly bent. When I had a head cold that prevented my sinuses from draining properly, I needed to sleep in a more upright position. On those nights I learned that the sofa next to the recliner in the sun room was most comfortable.

On nights when my immune system was low and my wife thought she might be catching a cold, I vacated the bedroom and slept in the sun room. On nights when my nose was clogged from a head cold and I couldn't sleep quietly without coughing, snoring, or constantly sitting up to blow my nose or clear my throat, I slept in the sun room. On nights when my medications had my mind racing and I simply could not sleep, I sat in the sun room.

During those days, the family knew I was sleeping in the sun room, so they were quiet and careful about turning on lights or being noisy in that part of the house. It was inconvenient, but it was necessary.

Pill Container

The weekend after I was diagnosed with cancer, I purchased a pill container with four compartments per day, Monday through Sunday. It became obvious very quickly that I would be taking a bazillion pills and that I needed a way to lay them out to keep accurate track of them.

The four compartments each day were labeled 'Breakfast', 'Lunch', 'Dinner', and 'Bedtime'. Those four compartments were sufficient for my daily doses. I did not need to take pills 'every two hours' or 'every hour.' In most cases, instructions said to take pills two or three times per day. Even when the pills were supposed to be taken 'every four hours, four times a day', the four mealtime and bedtime compartments sufficed.

Some days I had as many as 15 pills in a compartment. It would have been almost impossible to pull out all of the pharmacist's bottles at each meal to take the required pills. Had I tried to dispense pills from the bottles at each meal, I could had forgotten some pills, and it wouldn't have been easy to know or to monitor. With the week's supply in the distribution box, when I forgot to take some pills, it was obvious! The pills for a particular meal were still sitting there in their compartment.

The pill distribution box also allowed me to put out the pills once a week. When the doctor made changes to the prescription dosages during the week, I simply added pills to or removed pills from the distribution box to meet the new requirements.

Pill List

As I mentioned earlier, I printed a *current medication form* off the internet and I used it to keep a list of my current meds. Several times during the treatment process, I started with a blank form and re-listed my current meds. Since the meds changed frequently, this form was an easy way to keep track of everything. The fact that it was a single sheet of paper also made it an easy form to copy and carry to all doctor's appointments and to all new procedures.

Every time I visited a doctor or a clinic, they wanted to know my current medications. This list was very handy on those occasions.

Pill Bag

Soon after my diagnosis, I took a small, strong shopping bag and organized it to hold pill bottles. I made dividers out of poster board and a cover label to show which pills were in each compartment. Then, I filled it with pill bottles and boxes from the pharmacy. The bottles for every pill I am currently taking sit in that bag. The bottles for every pill I take "as needed" are in that bag. When I stop taking a pill, its bottle can be moved to another bag to avoid confusion.

Every Sunday evening or Monday morning when I need to refill my pill distribution container, I bring out both the pill box and this bag of pills. When I run out of pills, I phone the pharmacy to have those prescriptions refilled.

Each prescription is usually accompanied with a sheet detailing its uses, side-effects, complications, etc. I keep copies of all such paperwork in the pill bag.

I only need to use the pill bag once a week — on the day I refill the pill distribution box.

Anti-Nausea Meds

The anti-nausea medications that are administered prior to chemotherapy are excellent. Out of the six month treatment process, I was nauseous for only one day and I think I caused that myself by overeating the night before.

I carry a bottle of anti-nausea pills that I can use whenever needed. They can be taken in addition to any IV anti-nausea medications because they are a different type of drug. Early in my chemo treatments, it was recommended that I fill this prescription so I have these anti-nausea meds handy — **just in case**. Waiting until becoming nauseous to fill a prescription will slow and complicate the process of treating the nausea. It was a good suggestion to fill that prescription. On that day when I became nauseous, I had those pills available for use.

During the stem cell transplant procedure, they gave me a continuous pump (connected to my PICC line) that contained anti-nausea medication. If I felt sick, I could push the button on the pump and self administer 2 ml directly into my blood stream. It was like a "panic" button — if I began to panic from nausea, I could push the button. That pump was comforting to have available even though I didn't need to push the button. I knew if I felt sick, those medications would go directly into my blood stream where they could work almost immediately. I didn't need to swallow any pills and hope they would stay down long enough to be absorbed so they could perform their function. I just needed to push the button!

I remember my mother telling us back in 1999-2000 that the current anti-nausea medications were really good. At that time, she went through chemo treatments for a cancer and she didn't get sick at all. Before chemo treatments started, the oncologist told me the same thing. He predicted that I wouldn't get sick during chemo treatments — he was correct.

I cannot imagine going through chemotherapy years ago before they had good anti-nausea medications. That must have been awful! But today, if a patient's body responds well to all medications, nausea should not be a problem.

If nausea is a problem, tell the doctors or chemo nurses so they can change the meds. Nausea **should not** be a problem for most who undergo chemo treatments.

Complexity of Meds

My particular treatment required a complex assortment of medications administered both orally and by IV. Some of my medications were to be taken only during the week of IV chemotherapy treatments. Some were constantly being adjusted to produce the right effect. Some were stopped and started based on scheduled procedures. Some were stopped and started based upon measured blood properties. All of this added to the complexity of the already complex treatment.

Each detail person needs to be completely aware of treatment requirements, prescription changes, and medications taken. The detail person especially needs to know names, numbers, and times of meds taken "as needed" for various problems. All such information should be written into a notebook so the details are available to any attending physician who asks.

Side-Effects of Meds

With the complex assortment of pills and IV medications I was given, it seemed like every side-effect imaginable was possible. Some medications could cause totally opposite side effects (e.g., diarrhea and constipation.) This basically meant that some people might develop one, some people the other, and some people both side-effects.

Since manufacturers are afraid of lawsuits, they will list every possible side-effect on the paperwork that accompanies each medication. Although I started to read all of the possible side effects for each medication, I eventually gave up the practice. I just assumed that the new pill would mess up my system again, and generally — that was correct. The doctors and nurses warned me when there was an especially important side-effect that I needed to monitor.

For example, I was taking some pills to counteract side effects of other pills. In those cases, it was obvious that dangerous side-effects were possible. In most other cases, side effects were more of the nuisance variety and it was a matter of 'wait and see.' My bag of pills contains medications that fight nausea, diarrhea, constipation, stomach acid, and gas. I carry them all in the bag so they are available when and if needed. Most were carried for a long time, without ever having the need to use them.

Just be aware that lots of side-effects are possible from cancer treatments — some may occur and others not. It differs with each individual.

For example, before receiving a recent new IV infusion, I was handed a sheet to sign to give my permission. The sheet listed 20 or more possible side-effects. I needed to acknowledge that I had been informed that all of those listed were possible.

How Will Your Body React?

I hinted at this above. I suggest that you assume each new pill will mess up your system again — then wait to see what particular effects the new pill has had. Sometimes this happens when a new prescription is added to the many pills already being taken. Sometimes this happens when a particular pill is removed from the active list.

For example, one of the most recent pills added to my list is the "water pill" given to me to prevent rising BP. I looked at the side-effects list to see how strong or mild this pill would be. The side-effects listed included almost everything possible. It particular, it could cause diarrhea and constipation. My insides had just been beginning to feel normal again after more than a year of chemo, steroids, maintenance drugs, etc., and when I added this new pill — a particularly tiny pill — it sent my insides back into a state of uproar. Slowly, after several weeks taking this pill, my insides began to calm down again and return towards normal.

As another example, during each office visit following a cycle of the cancer maintenance drug, they would ask me how my body was holding up to the drug. The maintenance drug was to be taken each day for 3 weeks, followed by a week of rest. I was taking a steroid and a blood thinner with it. In that particular case, my body was fine on the days when I was taking the maintenance drug. It went berserk on the days between cycles when I was not taking it. After a little experimentation, I decided that it was the abrupt stoppage of the steroid that was causing the problem. So I continued taking the steroid through the next rest week and had no problems.

I was also taking a sulfa drug to prevent a form of pneumonia. I only took those pills on Saturdays and Sundays. It took my body a few weeks to become accustomed to those meds.

I have come to learn that each new pill added to the list of prescription drugs to be taken will probably mess up the delicate balance your body has achieved. I suggest you simply assume that this will happen and give your body several days to accommodate each new medication.

Eating, Smelling, Tasting, & Cooking

At different times during my chemo treatments, I was hungry all the time, or I wasn't hungry at all; some times food smelled good; sometimes, food didn't smell at all; and sometimes, it smelled bad; sometimes my taste buds worked; sometimes my taste buds changed the flavor of the food; at other times, my taste buds didn't seem to be working; and sometimes cooking was okay while at other times, just the thought of cooking made me feel sick.

Be prepared for times during chemotherapy when all such possibilities will occur. This is apparently par for the course — so be prepared.

Ports

Pictures of my 'temporary' ports were shown in earlier chapters. They sounded terrible when first mentioned to me, but each one served its purposes well.

The PICC line in my arm was inserted by a nurse practitioner in a hospital room. This was one of the first procedures performed after my diagnosis and it was somewhat scary. They stuck a tube through my skin and into a vein so they could administer medications and withdraw blood easily whenever necessary. After five cycles of chemo (and about 6 months duration), they removed it.

The suggestion that they may need to install another PICC line is not scary at all. If I must undergo more chemo, I won't look forward to it because it is inconvenient when it is in place. But I won't worry about it either. If I need one, I know it will make any new treatments go much easier.

The two major nuisances with PICC lines were that (1) the line (actually, my whole upper arm) had to be wrapped with plastic wrap and sealed with tape before I could take a shower and (2) the dressing covering the entry hole in my arm had to be changed weekly. The advantages of the PICC line were that they didn't need to stick me every day to find a vein from which they could pull blood or into which they could administer IV meds.

This one advantage outweighed the nuisances! The PICC line saved me from so many venipuncture pricks that I am thrilled that they put the line in my arm when they did. The fact that the doctors and nurses had easy access to my circulatory system was great.

Now that I no longer have the PICC line in my arm, every time they want to take blood or give me IV meds, they need to find one of my veins to stick a needle into it. That is easier said than done. On one occasion, they stuck me twice before they found a vein, and then, that IV didn't work. So they had to remove it and try again. With the PICC line in place, they had no such problems.

There is a second, more permanent port that also could have been used. Its tubes ultimately end up in the same spot in the large veins near the heart, but this access port is under the skin on the chest near the clavicle. This doesn't require constant dressings or dressing changes because it is under your skin. It does require that they puncture your skin each time they need to access this port.

I cannot speak from experience because I did not have this second type of port. I saw many people who had them. For chemotherapy treatments, a port is necessary — but nothing to be feared.

I cannot recommend one over the other. The doctor will know which will work best for each particular cancer treatment process.

24/7 Chemo with Pumps

My treatment called for 24/7 administration of IV fluids and medications during the weeks of chemotherapy. A port makes this an easy process to accomplish. On the first day of each chemo cycle, they connected the IV bags to the lumens of my PICC line. When finished, they connected the pumps and put them into a relatively large shoulder bag that I carried 24/7.

At bed time, I hung the bag over the bedpost or laid it beside me in the bed. The tubes were long enough to easily reach from my arm to the bag. To take a shower, I had to hang the bag over the shower curtain rod so it would not get wet. In this case, the tubes were short enough that I needed to hang on to the curtain rod during the whole showering process — which was somewhat inconvenient. The choice to take a shower or to wash at the sink, however, was mine to make — and with the 24/7 pumps connected, it was definitely easier to wash at the sink.

Weakness, Standing Up, Balancing, and Going Stairs

The weakness that I suffered during my chemotherapy treatments made it difficult to stand up, to sit down, and especially to go stairs. Whenever I needed to use my thigh and stomach muscles, the weakness was readily apparent. My balance also took a major hit during chemo treatments.

Frequently, while sitting, I would lean forward and rest my head on the handle of my cane. During those times, my wife would wonder what I was doing. My usual answer was, "I'm trying to get psyched to stand up." Sitting wasn't difficult. Walking (once on my feet) wasn't particularly difficult either. But standing up and sitting down were difficult!

Going **down** the stairs, from a **muscle-use** point-of-view, was easy. From **balance** and **sure-footedness** points-of-view, going **down** the stairs was difficult. If I lost my balance, my grip on the hand rail, or I stumbled, gravity would pull me all the way to the bottom. So I needed to go down the steps one step at a time, slooooowwwwwlllllyyyyy.

Coming **up** the stairs, was difficult from a **muscle-use** point-of-view. If I lost my balance, I would fall against the steps which was not a major concern. But raising my body one step at a time with my leg muscles was difficult. For quite a while, I used both arms and legs to go up each step. One arm pulled on the hand rail, one arm pushed on the cane, and the legs pushed me up to the next step. It was difficult!

As I became weaker and weaker, I cut back on the number of times I used the stairs each day. I was down to no more than one trip down and up the stairs per day by the time of my high dose chemotherapy. At my home, it was actually easier to walk out the back door, around the house, and up the sloped driveway to the first floor than to walk up the steps.

As my balance deteriorated, it became more and more difficult to walk a straight line without a cane. To stand still and have a conversation with someone, the cane was an absolute necessity.

During my first two-week stay at a hotel in Greenville, the hotel's only two-bedroom suites were all on the second floor. Each suite had an outside entrance and an individual flight of stairs. For the second stay during the high dose chemo and stem cell transplant weeks, I requested a two-bedroom hotel suite on the first floor. The coordinator found a different hotel that had two-bedroom suites on the first floor (no stairs!) That was great!

Be prepared to lose strength, balance, and the ability to go up and down stairs easily. I realized what I always took for granted regarding stairs when I watched my sons take a flight of stairs in about two seconds using only four steps. While I struggled one-step-at-a-time and blocked everyone behind me. Whoosh-whoosh-whoosh — they were up the stairs! Not me, though!

Be prepared!

23

Why Me???!!!

It may be common after a diagnosis of cancer to go through a period of soul-searching while attempting to answer the question, "Why Me???!!!" In my case, this did **not** happen.

The first Sunday morning speaker I heard after completing five chemotherapy cycles complimented me on my e-mail updates to friends which were all positive and up-beat. He said he benefitted from and was encouraged by my positive attitude throughout this whole process.

It had not occurred to me until his comments that I had not asked God this question throughout the whole chemo treatment process. I did ask God lots of questions and I requested lots of guidance, but I never angrily (or even calmly) asked Him, "Why me???!!!"

God Is In Charge

God Causes Events to Happen???

I know from experience that some people believe God controls everything — every little thing! They talk about His **sovereignty** — which is OK. According to them, nothing happens that God didn't **cause** to happen!

God **is** sovereign, but the problem with their extreme view is that it leaves no room for human sin. If a person has a few too many drinks, drives his car, steers in front of an 18-wheeler, and is killed, God didn't cause that! The alcohol did! The impaired driving did! The bad decisions to drive and to turn onto a busy highway did! The person himself or herself caused the accident! If God causes each such incident to happen, that point-of-view suggests that God sins, or He causes people to sin — and that is inconsistent with God's sovereignty in action.

God certainly **allows** things like that to happen, but that is totally different than saying that God **causes** everything to happen.

God Allows Events to Happen

God <u>is</u> in charge and He **can** cause anything and everything to happen, but in most cases, He doesn't need to **cause** things to happen — He just needs to **wait** and **allow** this fallen world to run its course.

After the fall in the garden, God didn't need to **cause** cancer in individuals — He only needed to **wait** until it happened because we live in fallen, frail bodies in a sinful, ruined world. Some people will get cancer. Some won't. Some will have heart attacks. Some won't.

God didn't cause me to have cancer — He allowed it to happen. I'm sure the question He was asking was, "How will Dennis deal with this?" It is obvious to me that God knew long before I did that I had cancer, and He was preparing my family and me for it long before my diagnosis.

God Wants Our Attention

When we are in trouble, God wants our attention. He wants us to draw near to Him. Whether we are Christians or not, He wants us to come to Him for help. He uses each of our follies to grab our attention, to draw us closer to Him, and to cause us to communicate with Him.

He certainly used this cancer to grab my attention! And I gave Him my full attention. I wanted to know what He was trying to tell me, what he was trying to teach me, and what He had in store for me. I also wanted, of course, for Him to help me get past this.

At no time during this ordeal, however, did I ever shout at Him in an angry voice, "Why did You do this to me!!!!???" I know that some people do react this way. But I also know that God loves me! I don't believe He would intentionally give cancer to me any more than I would intentionally give cancer to one of my children. Now that I have it, I know He is there to help me battle it.

As a result, I did ask Him over and over again what He wants me to learn through this experience, how He wants me to change if I live through it, and what services He wants me to perform post-cancer.

Near the end of my chemotherapy treatments, the same nurse who told me God would have a work for me to perform after battling cancer told me, "After you have finished the treatments, God will put lots of people in your path who can use your help."

God wanted my attention and He got it! Since my diagnosis, I have been paying particularly close attention to Him and to His guidance. To whom else can one turn when entering, enduring, or surviving such a battle?

God Wants Us to Grow

God wants us to grow as we go through life. This cancer episode has caused me to learn many things which were totally foreign to me. Looking back, I must admit I learned lots in the physical, emotional, and spiritual realms.

Physical

I have always known how strong and resilient our physical bodies are, but as a result of this bout with cancer, I learned how weak and fragile our bodies really are.

Even now I watch as my kids bound up and down the stairs. I marvel at how easy it is for them, and I compare their abilities to mine — before, during, and after cancer.

Before cancer, I could bound up and down the stairs, too. Immediately after my cancer treatments, going down the stairs required one step at a time, slooowwwlllyyy, so as not to lose my balance. One day, for whatever reason (????), half way down the stairs, I wondered what type of light bulb was lighting the stairway — so I looked up and promptly lost my balance. Fortunately, I had a tight grip on the hand rail, so I was able to control it. Most people hardly ever give balance a second thought. We can balance, walk, jump, run, etc., and we don't need to think about any of it. But since chemo treatments began, I have had to pay daily attention to lots of things.

Before cancer, I could walk in the dark or close my eyes, and not lose my balance. Now, I can lose my balance when washing my face with my eyes closed in the shower. Now, I make sure I am leaning against the wall or holding onto something before I close my eyes and wash my face in the shower. I never had to give it a second thought before.

Before cancer, my fingers, hands, toes, and feet worked well without thought. Now, they hurt much of the time and they don't move well, properly, or as expected. Now, I must intentionally pick up my feet higher than before when I walk because my heels rise on command but my toes don't necessarily do so. If I shift to the right without thinking about it, the toes on my right foot drag, my foot doesn't necessarily move to where I think it did, and if I shift my body's center of gravity to the right — BOOM!! I can find myself down on the floor.

I did this once in the hospital. Talk about instant excitement!! Every nurse in that end of the building was surrounding me in about 2 seconds. I was facing one way, wanting to slide my foot to the right, while talking to a nurse out to my left ... and BOOM!! My heel moved, but my toes didn't, so when I shifted my weight, I went down! I always joked that some people can't do two things at once — (1) walk and (2) chew gum. Well, I can't walk and do much of anything else at the same time now. I need to concentrate on what I'm doing — one thing at a time!

I never thought about any of this before cancer. I now know that we are fortunate that our bodies work as well and as wonderfully as they do for most of our lives — without the need for us to think very hard to do much of anything. We take so much for granted — but post-cancer, all is changed.

Emotional

In the past, I have always been able to keep my emotions in check. I noticed that after my father-in-law had surgery and chemo for a cancerous brain tumor, he became much more emotional than ever before. Now that I have experienced cancer and chemotherapy, I have become much more emotional than ever before, too.

It is even difficult, sometimes, to pray without crying. Depending what I'm saying or requesting, I can't do it.

When friends tell me they are praying for me, I frequently can't make a sensible reply to them — because I'm choked up that someone would consider praying for my family and me. A good friend gave me a ride home one night and told me as I was ready to get out of the car that he and other friends were all praying for me. I couldn't talk. I stepped out of the car with a huge lump in my throat, and mumbled something back to him. Several days later I told him how moved I was and how much I appreciated his comments — but that I was so choked up I couldn't speak or say anything at that moment.

One day recently, the pharmacist saw me and came out from behing the counter. He put his hand on my shoulder and said, "You look great. God is good!" I was crying before he finished and I struggled to respond.

I have learned a lot during this time as my emotions kick in regularly and strongly. I don't know how different I seem now to others, but my emotions now seem very different to me.

Spiritual

I think I am spiritually closer to the Lord post-cancer than before. Spirituality isn't something you can look into a mirror and assess, but I think I have changed spiritually, too.

For one, we had to rely on the Lord for our livelihood during chemo treatment days in particular. My wife has an hourly job which mostly pays for the health insurance. Beyond that, her checks contribute little. Over Christmas of 2008, when chemo treatments were in full swing, I was disabled (with no income) and she was laid off for three weeks. Our income was essentially zero — but we had a wonderful Christmas! — because the Lord provided.

My wife called attention to the fact that during that time, we didn't miss paying any bills due to lack of funds. Even though I am totally disabled and eligible for Social Security disability payments, the Social Security Administration has a five month waiting period before they pay **any** benefits. Then their payments are on a one-month delay. Even though my date of disability was in early June, 2008, I wasn't eligible for any disability payments until late in January of 2009. So we went six months with only my wife's meager income, but the Lord provided!

We didn't become rich during that time, but neither did we lack. We had what we needed when we needed it! The Lord is amazing, isn't He??!!

During that same time, the Social Security Administration was deciding whether or not I was disabled. It is a blessing that they made a quick decision in my favor. According to TV ads, they are not always so accommodating.

I also learned during that time, and had it reinforced, that the Lord **is** in control. He can step up to the plate when He wants to! In one specific event, He prevented me from making a bad decision by taking a decision out of my hands. I know that my family and I are in good hands — the Lord's hands. We continue to look to Him, and I know He is providing for our needs!

Before cancer, I was much more self-assured and independent. I always said we were relying on the Lord, but with a substantial monthly salary check to support us, our reliance on Him was really only token. Take away the substantial monthly checks — and we learned to really rely on Him.

I can only hope that these changes in my thinking become permanent parts of my life — because I like it when I can say that this verse expresses our attitude: "**Casting all your care upon Him; for He careth for you.**" (1 Pet 5:7) I may not always have lived that way in the past, but I want to do so in the present and future.

Thankful

If I didn't angrily confront God with a "Why me???!!!!", what did I think or say to Him? Looking back, I realize I was quite thankful that I was the member of our family who had cancer. It wasn't my wife or any of my children. I feel badly enough that my wife has the physical ailments she does — it would have been much more difficult for me to watch her go through this than for me to go through it myself. I am thankful and full of praise that the rest of the family is healthy and well.

I am thankful that my body has responded well to the chemotherapy treatments. I am thankful that God has been on my side throughout this battle.

It Would Have Been Easy ...

I never angrily confronted God and I don't plan to do so. Looking back, however, I can see that it would have been very easy for me to accuse God of doing this to me, to be angry with God, and to just roll over and give up. I can see Christians as well as non-Christians doing this.

Of all people, Christians should not respond angrily towards God because they look forward to eternity in His presence. But Christians are not perfect — so some of them might choose to respond negatively.

Without a close relationship to God, it would be easy for anyone to respond negatively to a diagnosis of cancer. But God doesn't want a negative, angry response to be the end of the story — He wants everyone to turn to Him, to return to Him, and to move closer to Him! He is ready and willing to help everyone! His offer of forgiveness, comfort, guidance, and blessing is to everyone — not to just a special few people — but to EVERYONE!!!

When we each recognize we need His help, His arms are open wide — He is ready and willing to welcome each and every one of us into His fold!

I expect my positive attitude to show through this book. Don't look for me to be angry with God. I wasn't angry with Him at diagnosis, during treatment, or now. Do look for me to be up-beat, positive, and optimistic! My future in service to the Lord is within His control and I anxiously wait to see what new tasks He has in store for me.

If you haven't one already, each of you reading this book can have a similar confident, positive attitude towards your future. The Lord Jesus Christ is a loving God Who wants everyone to draw near to Him in His fold. **"The Lord is not slack concerning His promise, as some men count slackness; but is longsuffering to usward, not willing that any should perish, but that all should come to repentance."** (2 Pet 3.9) Today is a good day to take advantage of this!

24

A Supportive Spouse & Family

I realize some multiple myeloma patients may not be married, so for you, the title of this chapter should be translated, "**A Supportive Family & Friends**". I am married so I will write from that perspective. Translate this chapter to meet your own personal situation — but don't ignore it! You may be a Christian and have the Lord on your side, but you still need lots of support during this time from spouse, family, and friends! You cannot do this alone.

Extremely Stressful Days for All Involved

Scheduled Medical Procedures

Going through chemotherapy and a stem cell transplant is a very stressful time. You may not notice the mental and emotional stress at the time, but it is there. The most obvious stress is physical stress. Extreme weakness plus the necessity of all of the medical procedures (many of which entail pain) are physically stressful. Pain and stress may fade with time, but even when procedures are painless, stress is present.

At the start of this whole process, I had seen few enough doctors that my blood pressure probably rose about 20 points just at the thought of having to visit a doctor. The same happened at the thought of having a vein punctured to draw blood. Etc. Now all such visits and events are second nature to me. (Ho, hum, yawn!)

The thought of having temporary catheters and ports surgically implanted or having to endure more bone marrow biopsies still produce lots of stress. Emotional and mental stresses precede and follow many such procedures while physical stress accompanies surgery and maintenance treatments.

I entered the realm of the unknown when I had my first bone marrow sample taken. That event was accompanied by lots of anxiety and stress. I had heard they were painful, but I really didn't know. Then, I found out that they use local anesthetics, so I relaxed a little. Then I learned from experience, that even with local anesthetics, bone marrow biopsies are painful. (Live and learn!)

Before the second and following bone marrow biopsies, I asked for more pain killers. The first sampling procedure was educational — I learned bone marrow sampling is painful and stressful. As soon as the procedure itself was over, however, the pain killers did a wonderful job! After several more bone marrow samplings, experience teaches that they are painful and stressful, even **with** the administration of extra pain killers.

I get anxious thinking about the unknown. I know this. Port insertions, bone marrow biopsies, and colonoscopies were all unknowns when first mentioned. Now that I have experienced two catheter insertions (and removals), multiple bone marrow biopsies, and a colonoscopy, I am much less stressed at the thought of more such procedures. Although one might think so, the stress never completely disappears regardless how many such procedures one has experienced.

My friends and family know from experience that I am most anxious about bone marrow biopsies. They hurt — I don't like them — and everyone knows in advance when I am scheduled for another — because I let everyone know I have one coming up!

Your spouse doesn't necessarily know how such procedures feel (unless having experienced one for herself at some point in time) but she can see how the procedure, and anticipation thereof, affects your reactions, anxieties, and behavior. I am sure that when my wife sees my anxiety in advance of bone marrow samplings, she becomes stressed, too.

This occurs every time such a procedure is planned. In addition to ports, biopsies, and colonoscopies, the list of stressful procedures can include chemotherapy sessions, stem cell collection procedures, transplant procedures, scheduled periods of housing away from home, doctor visits, etc. Practically everything done during the whole cancer treatment process can be extremely stressful on you **and your spouse.**

So be aware and be mindful. You need a supportive spouse, but stress can and will take a toll on **both** of you — even though only you (and not your spouse) are directly subject to all the procedures.

Unscheduled Sicknesses and Complications

There are gigantic stresses associated with unscheduled complications that arise during treatments.

After my Cycle #1 of chemo, which hardly affected me at all, there was one day when I had both nausea and diarrhea. I had no energy that day whatsoever. Running 50 feet to the bathroom with zero energy was very nasty. Standing up was almost impossible. That day was also very stressful on family members who were trying to manage the process while providing the care I needed.

My focus that day was on myself — totally — my stomach, my intestinal tract, my nausea, and my total lack of energy. One of my sons was my primary care giver until my wife returned home from her own doctor's appointment. When she called to check in, she said she could tell by the sound of my voice that something was wrong — and she came straight home. She was immediately worried (stressed) about me and I know I didn't notice nor appreciate **her** level of stress. Looking back, it had to be intense for her and for my son. I am fortunate that my son was home with me that day and that my wife came immediately after her phone call. I am fortunate they both were very patient with me. I know this now. At the time, I was unaware.

After Cycle #2 of chemo, I came down with a mild infection at the same time that I had very low blood pressure (BP). The two may have been from different causes, but they occurred simultaneously. I learned that day that when my BP is below 100 (i.e., in the vicinity of 95/60 or less), I am light-headed with little balance and no strength when trying to stand up and walk. Sitting is OK. Standing is not.

That day, my BP took a dive at the same time that I had a slight fever. I remember feeling very weak. The chemo nurses told my wife to bring me directly to the clinic — which was easier said than done! Going to the clinic required that I walk out to the car. If I couldn't make it to the car, they told my wife to dial 911 and request an ambulance. Again, I learned after the fact how stressed my wife was! I knew that I felt totally lousy, I had no energy, and I had trouble walking. I was stressed! My focus was on me. My wife's focus was on me. But I was unaware of her stress — she seemed perfectly calm and collected.

With her help, I made it out to the car. She drove me to the clinic and brought me a wheel chair. When they saw me being wheeled into the chemo room, they decided almost immediately to admit me to the hospital. They told my wife to drive me straight to the hospital and take me straight to my room. That 50 minute drive passed without incident. Then, we both relaxed substantially when I was in my room and I was under the care of the doctors and nurses at the hospital.

The fact of my cancer has been stressful to everyone in my family, but events like this continue to cause everyone's stress levels to peak on a weekly (and sometimes even daily) basis. I know this now. At the time, I was oblivious to it.

Strong Support from Your Spouse

Had my wife been the one with the cancer, I don't know how well I would have filled the support role for her. I hope I would have done as well for her as she did for me — but I don't know. (And I hope I never need to find out.)

At the time of my diagnosis and during my treatments, I must confess that my thoughts and concerns were primarily on myself. I know my wife is strong. But during this process, she had to be strong for me without much support from me. She, nevertheless, has been extremely supportive throughout this whole experience and I am totally grateful to her for it.

Throughout this ordeal, I know that my wife has been dealing with her own physical problems but she has kept quiet about them. Most of them were unknown to me. My focus was on my problems and she said I didn't need to be additionally burdened by her problems.

It is extremely important that the person with cancer has a strong, supportive spouse throughout the whole treatment process. I have been blessed to have the support of my wife throughout my battle.

My wife knows I have total confidence in my team of doctors and nurses. As long as I am able to make decisions for myself, she knows I will take their advice in most cases. She also knows how I think. So when I am not able to make my own decisions, she will make good decisions for me.

For those who are single, the major support person may be a parent, an older child, or a close friend — whoever is designated to make decisions in the event of your incapacity. Someone must fill this role and it is best to decide early in the process who that person will be. It **MUST** be a trusted individual! — one whom you trust implicitly!

My wife and I each have *durable* powers-of-attorney for each other, so we can make decisions and sign papers for each other when necessary. Many single people haven't thought or planned that far ahead. Such a power-of-attorney is something for each young person to consider obtaining as they leave home to begin life on their own. Just remember — if you don't clearly state your intentions, or designate who will make decisions for you in the event of your incapacity, the state can and will make those decisions for you.

What Makes A Supportive Spouse?

You might be thinking that some spouses will be supportive and others may not. Spouses can and should play supportive roles through times like this. What makes a supportive spouse?

A supportive spouse

• is one who is there for you under any and all circumstances.

• recognizes the physical, emotional, and mental burdens that are occurring — that is, they recognize that you will not act normally.

• attends as many doctors appointments as possible, listens to the doctor's comments, recommendations, opinions, and directions, feels free to ask questions, and after the office visit, discusses and rehashes everything with you in detail.

• knows how and why you think what you do and can substitute for you as decision maker when necessary.

• understands the need for a bazillion medical appointments and procedures and helps to make them all possible by getting you to each one on time.

• is one who can jump in and insure that you are taking all your medications in proper dosages at their proper times.

• is one who can offer any kind of support whenever and wherever necessary.

When no other tasks are required, a supportive spouse

• will sit by your side in a hospital room — if for no other reasons than to hold your hand, to keep you company, and to be there for you.

• etc.

As the cancer patient, you need to recognize that you are causing your spouse, family, and close friends untold amounts of stress. There is little you can do about this, but if you recognize it as it's happening — you will be a much better and certainly a much more appreciative spouse, family member, friend, and cancer patient.

After Your Spouse Come Family and Friends

It would have been impossible for my wife alone to do everything needed to support me throughout this whole process. On many days of chemotherapy and hospital procedures, I was not allowed to drive myself, even though I was capable of driving to most appointments.

My wife's employment was providing our health insurance coverage, so she could not take off at any and all times to drive me places. According to federal law, she could have taken several weeks off to care for me and have a job waiting for her when she returned. There were two problems, however, with taking advantage of that law: (1) The law guarantees a job with the same company but not necessarily **the same** job; and (2) Leave taken under this law is unpaid leave — so time off consistent with this law produces no income and no insurance coverage.

In addition to my wife, I also relied on the help of family and friends during my treatment process. Lots of friends volunteered to help "in any way needed." I took advantage of some of their offers when I needed someone to drive me to appointments. The cancer clinic closest to home is only about 10 miles away which is a relatively short drive. Many of the procedures, especially transplant procedures, were performed at the clinic in Greenville, SC, which is about 40 miles away (in the opposite direction). All procedures performed in Greenville required larger chunks of time than procedures at the local clinic — simply due to driving times.

Twice during the treatments, I needed to move to a hotel in Greenville for two weeks. The first time was in preparation for and during stem cell collection procedures. One of my sons, a grad student at the time, had the most flexible schedule. He functioned as my primary "care giver" during much of that time. My wife filled that role on weekends and as much as possible during the rest of the two weeks (work permitting).

The second trip to Greenville was during the week of high dose chemotherapy and the stem cell transplant. In that case, one of my nephews stayed with me. Again, my wife stayed with me on weekends and as many other days as her work permitted during those two weeks.

Other friends graciously drove me to and from many appointments. Sometimes, one person dropped me off at the hospital or clinic and another person picked me up. When it became complicated, we worked it out.

In all cases, when I needed help, the help was there. I am blessed to have had such support during that time.

The Necessity

I am blessed to have the support of a wonderful wife who loves me and who is concerned for my best interests. I consider such support a necessity when one must go through the cancer treatment processes.

For single people, it is necessary that someone fills this role. For a young person, a parent can do this. For a parent, an older child or a close friend can do this. Someone needs to be there for you at all times. For example, when my mother was being treated for her cancer, she lived with us and my wife took on the roles of detail person and primary care giver.

Whoever fills the role MUST be supportive of the whole process. Someone who thinks the patient or the doctors are going about treatment in the wrong way should not fill this role. Spouses, parents, or children in this role should agree with the patient's choice of doctors and treatment processes.

Finally, you **MUST** remember that the whole process is stressful for all involved — especially for the primary care giver. Every family member will be stressed. Spouses will be particularly stressed. None of this may be apparent to the patient at the time, but stresses **will be** present. Keep that in mind and appreciate how blessed you are to have spouses, family, and friends supporting your every step during the treatment process.

25

The Lord's Long-Range Plan for Us

Chapter 6 presented a description of events leading to my diagnosis with cancer. For those of you who are interested, this chapter presents a description of events of the past twenty years that led to our being here in Clemson, SC, where my diagnosis occurred. The whole story needs this longer-term chronology that led to us being where we are today.

Academia

We are here in Clemson because I am a ceramic engineer. I am currently an Emeritus Professor of Ceramic and Materials Engineering at Clemson University. Over the last two decades of the 20th Century, I taught on the Ceramic Engineering faculties at both Alfred University and Clemson University.

I became a professor because I wanted to teach. Until the 1970s, that was the primary responsibility of most professors — they taught. Some did research, but many spent their whole careers teaching undergraduate and graduate classes. Then things started to change. I joined the faculty at Alfred University in the early 1980s when the emphasis had begun to change from teaching to research.

My father-in-law (a Professor of Ceramic Engineering at Alfred University) and I had worked together for several years. We decided we would each go further if we worked together than if we tried to work individually – so we became a team. The vast majority of our publications throughout the 1980s and 1990s listed both of us as co-authors. We wrote a large ceramic processing textbook together. We did continuing education seminars together. Our names are still linked throughout the ceramics industry.

When father-in-law announced he was going to retire and move south, he had already talked with the Head of the Ceramic Engineering Department here at Clemson University. He told the department head that he and I were working together, and we would like to continue to do so, if possible. After hearing that, the department head handed him a job announcement for a faculty opening and told him to give it to me. If we were both here in the area and I was on the faculty, we could continue to work together to the benefit of all. Next thing I knew, our family was moving to Clemson, South Carolina, and a year later, the in-laws moved down here, too.

The Ceramic Engineering Department at Clemson was a small department with lots of courses to be taught, so I was doing exactly what I wanted. I was teaching and developing courses in my specialty. After I arrived, Clemson, too, began to emphasize research. I was successful at arranging a fair amount of industrially sponsored applied research, so I successfully moved up the academic ranks to Professor.

Obsession with Research

Then, the administration began to obsess about research. It didn't matter how old you were, what academic rank you held, how many courses you were teaching, or anything else. They expected you to be doing lots of research. I was successful at bringing in industrial funding for applied research. The government, however, wasn't funding any research in my area of expertise. Even in the 1990s, they considered traditional ceramic processing technology to be a *mature* technology — *mature* technologies require no governmental research or research funding.

In the late 1990s, I had begun consulting in Indonesia. I traveled back and forth to Semarang, Indonesia for several years, including a sabbatical year. I decided to make this my primary job, so when I returned to Clemson following the sabbatical, I announced that I was retiring at the end of the school year.

If they want professors to do mostly research, they need to hire research professors. They are doing that now and research expenditures are booming! But now they are struggling to find good teachers, because most of the good teachers have retired. Go figure!

The Lord's Involvement

The consulting opportunity in Indonesia appeared to be a door that God opened for me. They don't have ceramic degree programs in Indonesia. So every day in Indonesia presented teaching opportunities to everyone from the owner of the company down to lab technicians, plant engineers, and workers. The boss was educated in America as an engineer, but not as a ceramic engineer. None of the engineers in his plant ever studied ceramics. None of the staff or plant workers ever studied ceramics. They all learned ceramics by OJT (On-the-Job Training). So teaching opportunities abounded! It was great!

This consulting opportunity had been plunked into our laps. One day in the mid-1990s, my father-in-law came to me and said, "Pack your bags, we're going to Indonesia." At that time, I didn't even know where Indonesia was, but we were going there — for a week. God gave us this project; we were going; and the next thing I knew — it was all arranged! This was a pretty good sign that God was telling us, "Go work in Indonesia." So we went.

After that first trip which father-in-law and I made together, we each continued to make four trips a year to Indonesia for the next several years. Most of the time, he and I went separately. We tried to overlap a little at the beginning or end of each stay — the goal was to have only one of us in country at any one time. The Indonesian staff could handle one of us at a time. Two at a time overwhelmed them.

Father-in-law's trips came to an end in 1999 when he was diagnosed with a malignant brain tumor. My trips continued for several more years.

Sabbatical Year Goals

During my sabbatical year from the university, I had set several goals. My **first** goal was to work for a traditional ceramic production company. The company in Semarang produced both sanitary ware and tableware (china). The two factories were side by side in a single huge complex. The technical staffs of the two companies shared the same office and laboratory, so we all worked together. It was the great, **ceramic** industrial experience I had not had as a new engineer after graduating from Alfred University in 1970 or after Penn State in 1975. After graduating from Penn State, I worked in industry, but not in a traditional ceramics industry. In the Indonesian plant, I was involved with every staff member and practically every engineering project — it was wonderful! The staff was great! The country was great! And I have many fond memories of those days!

My **second** goal for my sabbatical year was to learn if I could live half way around the world for weeks at a time away from home, and whether the rest of my family could live here at home without me for the same length of time. We learned that we were all capable of this. One good thing about being away for one month out of three was that I was home the other two months out of three. I enjoyed that schedule. The part of the trips I didn't like was the 44 hour travel time on each end of each trip. Leave home Thursday morning; leave New York JFK Thursday evening; fly for about 24 hours; and arrive in Indonesia early Sunday afternoon (a total of 44^+ hours later.) Coming back, leave Indonesia on a Saturday afternoon and arrive home Sunday evening (44^+ hours later.) The long trips were tolerable, but the time in Indonesia and the time at home between trips were great compensation for the inconveniences and discomforts of the travel.

The **third** goal was to learn if they wanted or needed me in Indonesia for longer than just a few short years. That answer appeared to be a, "Yes."

So overall, it appeared that the Lord had plunked an opportunity into my lap to move on from Clemson University to follow the field of ceramic processing to its new shores. The US government wasn't interested in supporting research into what they considered *mature* ceramic processing technologies; most of the ceramic production industry was moving out of the US to the countries south and west; so in most cases, to work in a good ceramic production company meant to work in Mexico, Central or South America, or somewhere in the Far East.

Indonesia is a beautiful country, and with all this information, the Lord appeared to be standing there saying, "Here – the door's open. Step through." I did. So I retired from the university to become a full time consultant with one major job – in Indonesia.

Consulting Days

My schedule after retiring from the university was effectively two months at home followed by a month in Indonesia – four trips a year. This continued for several years.

I had a great time in Indonesia. One of my Ph.D. students took the job as the company's laboratory manager. She was there full time for more than 10 years. My brother-in-law, a ceramist and PhD chemist also moved his family there for several years. The Indonesians were great! Teaching was always part of the job. Problem solving throughout the plant was a daily practice. I have many fond memories of those days. I wish I could return and continue.

During the two months at home between trips, I spent my time writing books. I authored three paperback technical books during that time: <u>Particle Calculations for Ceramists</u>, <u>Rheology for Ceramists</u>, and <u>Character-ization Techniques for Ceramists</u>. I also authored a set of add-in functions that performed particle calculations in MS Excel®.

I also studied and wrote about Christian topics, and I edited two books transcribed from Jim Funk's Sunday morning messages. Some of these are available in paperback form; some are available only in electronic form. I kept busy studying, writing, and programming during those days.

Eventually, the job in Indonesia began to wind down. It became obvious to me that the owner wanted me to change the nature of my participation in his company in a way that didn't interest me. He wanted me to make more trips each year, but stay for shorter periods of time. That would have maximized the parts of the trips that I disliked — the time spent traveling. So the consultancy in Indonesia ended.

I continued to study and write as I advertised as a consultant here in the US, but consulting opportunities in my field in the US were few and far between. I had one or two good consulting jobs a year during that time, but that was it.

Advertising in professional journals, I learned, was expensive and produced no results. My brother suggested I write an electronic magazine (an "e-zine") and distribute it throughout the ceramics industry. I did that, and to my surprise learned that it was really inexpensive advertising.

My distribution list for the *Ceramic Processing E-zine* remained between 700 and 900 subscribers for several years. The e-zine was a free distribution, each of which contained an article on a practical aspect of ceramic processing. I received many compliments on those efforts.

A free distribution doesn't put food on the table, and the book sales (advertised in the e-zine and on my web site) never sold in sufficient numbers to put food on the table either. So I needed something else.

Oconee Christian Academy

At that point in time, both Chris and I were searching for jobs. We figured that if one of us could land a job with health care benefits, the other could remain home and finish our son Joe's home schooling. That was our idea anyway. But neither of us were finding anything. Chris had never worked outside the home, so finding and taking a job was all new to her. I was selective about the types of jobs for which I would apply. Actually, it appeared to be a good time for me to switch fields to get back into computer programming which I had always liked. So I searched for and sought jobs in the software development industries. Chris had no success. I had no success.

Finally, in August of 2005 while sitting at the computer, I remembered that I had promised myself to inquire at the local Christian schools. School mates of mine from Alfred University were involved with one of the local Christian schools, and I knew there was another Christian school in Seneca (the next town west of Clemson).

My problem had been that I'd make mental notes to search here or there with this company or that, but when I was sitting at the computer, I didn't remember the mental notes. That day in August, however, I finally remembered to check out the local Christian schools. I searched the internet, sent e-mails, and received an almost immediate response from the Director of the Oconee Christian Academy.

Turns out, about five minutes before my e-mail arrived, the principal had reminded the director that they still needed a science teacher. They were praying for the Lord to handle it. They don't advertise for the people they need — they pray about every need and turn all needs over to the Lord. So when my e-mail arrived a few minutes later, the director called the principal back into his office and said, "You've got to see this!"

I was invited to visit the next day. What could I teach? I told them I could teach "physics, calculus, algebra, physical science, earth science, and chemistry." What did they need the science teacher to teach? They told me he had to teach "physics, calculus, algebra, physical science, earth science, chemistry, and biology." So the only course they wanted me to teach that I really hadn't listed was biology.

An additional benefit of teaching at OCA was that each teacher received tuition credit for one child. Our youngest son, Joe, had been home schooled since his elementary school days and he longed for a larger circle of friends. He had some good friends through the home schooling organization, but most of those friends lived several towns and 30-45 minutes away from Clemson. So if I took the job at OCA, Joe could go to school there and join their 9th grade class — the Class of 2009.

OCA offered no health care plan, however, and as a Christian school, the pay scale for even a PhD with many years of teaching experience was not very great. Chris could continue job hunting; I still was making some money from book sales; and I could continue to consult. What to do?

The fact that the courses they needed taught and the courses I could teach matched so well, plus the timing of my e-mail, and the start of classes about two weeks later all suggested that the Lord had guided me to them. I would be teaching again; Joe would be able to join a reasonably large class; and Chris remained free to continue to search for a job.

We decided that this was the Lord's will and I took the job.

A few days after school started, Chris checked with the local food service company on Clemson campus, and they immediately offered her a job. Their prior employment requirements were minimal, but most of their jobs were physically demanding — requiring employees to stand on their feet for long hours. Fortunately, they provided good health care coverage for all full-time workers — and Chris was hired for a full-time position.

Our status at that point was this: I was teaching at a Christian school (which fulfilled my desire to teach); Joe was in the 9th grade at the Christian school (which fulfilled his desire to have lots of friends); and Chris was working a food service job (which fulfilled our desire for health care — but did nothing for Chris' desire to remain only a housewife). Since Chris' job provided health care, I was able to drop the expensive health care plan we had through my consulting company. The amount Chris had to contribute for **excellent** health care was well below the amount I was paying for **poor** coverage. The Lord had solved a whole bunch of our problems in one short week.

Before Chris took her job, in an effort to keep our health care costs low, we had a health insurance plan with a $10,000 deductible. We would be protected if one of us was thrown into the hospital for any length of time, but everything short of that would be on our own nickel.

That first year of Chris' new job, her deductible was about $25 and her health care coverage paid 85% of the rest. After reaching an out-of-pocket maximum, the insurance paid 100%. The actual number comparisons were mind-boggling. Through my group health care policy, I was responsible for at least $25,000 out-of-pocket each year (monthly fees plus deductible) before my health care coverage kicked in anything. Through Chris' health care policy, we were responsible for a yearly maximum of about $5,000 out-of-pocket (monthly fees plus deductible) and we began to benefit almost immediately.

During that first year on Chris' policy, adding together her premiums, co-pays to doctors, costs of prescriptions, and carpal tunnel surgery (on Chris' hand), our **total** medical expenses for the year were less than ⅓ of my consulting company's **premium** alone. Like I said, the numbers were mind-boggling.

It appeared that the Lord had stepped in and solved our health care cost problems. I was certain the Lord wanted me to be teaching at OCA, Joe to be in a class there, and Chris to be providing health care. Only two problems remained: cash was really tight, and Chris was working outside the home. If the Lord was really in charge, He would make sure we had money when we needed it. But why was Chris working? That answer remained elusive.

"You Are An Answer to Our Prayers!"

To determine whether or not you are doing the Lord's will is a difficult task. It is sometimes not very clear. To walk through OCA's doors two weeks before classes begin and find an exact match between your talents and capabilities and the school's needs adds a little more certainty to the answer. Coincidences like that don't happen often. To learn, after the fact, that no other science teachers inquired about the job added a little more certainty. But when a mother came up to me and said, "You are an answer to our prayers!" — that makes one pause to consider.

No one before had ever told me I was an answer to anyone's prayers, but I heard that more than once at OCA. The specific goal of one of the seniors was to attend the US Naval Academy. To do so, he needed to take calculus and physics. Without a science teacher, that was going to be a major problem. His mother said they had checked all other options (local tech schools, colleges, etc.), and nothing was working out. Without a new science teacher, he would have a problem! Then I arrived.

That was pretty heady stuff! For once in my life, I was certain, beyond any shadow of doubt that I was doing exactly what the Lord wanted me to be doing. I was happy. Joe was happy. But — Chris was unhappy.

Why was it necessary for Chris to work a difficult, crumby, tiresome job which she didn't like to allow me to do the Lord's work in a job that I did like? That question remained a big question for at least three years of our lives. Neither of us knew the answer; both of us were anxiously waiting to learn the answer; and both of us continued to seek other employment.

One good job for me in the field of ceramics would have solved the whole problem (or so we thought!) One good job in the software industry would have solved our whole problem. A better, easier desk job for Chris would have reduced the problem. But nothing was happening. Résumés were sent out but no replies came back. Chris had a few job interviews, but no offers. One job I thought would be ideal for Chris was filled before her résumé crossed the supervisor's desk.

All advertised faculty slots at both American and foreign universities were for professors who would develop "dynamic research programs." Teaching abilities were hardly ever mentioned in academic job postings. Administrative positions were a step too high for me — I couldn't demonstrate a proven track record: I had never been department head at Clemson, so I did not have that background listed on my résumé. I was generally overqualified to take industrial positions, or I lacked an appropriate industrial track record. Who would hire a retired engineering professor to be a plant manager? Lots of inquiries and applications were mailed, but no interest was ever expressed by anyone in return. Why?

Chris and I had concluded that if we were actually doing the Lord's will, no doors would open. If I was doing what He wanted me to do, why would He allow me to change jobs? I thought I **was** doing the Lord's work, so why would He allow me to change jobs?

I considered another point-of-view also. Maybe the Lord was punishing us (me) for a bad decision. Maybe we (I) had acted independently of His guidance. If I had made a mistake somewhere along the line, however, the Lord should be punishing me, not Chris. It just didn't make any sense. **Nothing** made any sense.

Fortunately, neither of us was upset enough to throw a tantrum, quit our job, and force the issue. We both continued as well as possible, trying to patiently wait to learn the Lord's will. I had the easier experience. Chris had the more difficult experience.

A Mistake to Retire from Clemson?

One nagging question during this whole time was: Did I make a mistake when I retired from Clemson University? Did I do exactly what I wanted to do at that time and justify it in my own mind as the Lord's will? ... or did the Lord actually want me to retire from Clemson when I did? Even today, I do not have a clear answer to that question.

I also considered that question from another point-of-view. If the Lord wanted me to teach at OCA, how would He accomplish that change? Even if I had known about OCA when I was working at Clemson University (which I didn't), it is highly unlikely that I would willingly have taken a 75% cut in pay and dropped health insurance coverage to take a teaching job at OCA. It simply would not have happened.

Then again, it appears they needed the science teacher the exact day I enquired at OCA, not several years earlier when I retired from Clemson University. So the timing would have been off. I believe the Lord wanted me teaching at Clemson University, and then a few years later He wanted me teaching at OCA. Did He orchestrate it? Or did I make a series of selfish blunders from which He rescued me to move me from Clemson University to OCA? I still don't know.

Have I led a perfect life, 100% in tune with God's will? Certainly not! Had God been punishing me for making such blunders? Certainly possible! But at OCA, I liked what I was doing and I knew I was fulfilling the Lord's will for me. That was not punishment.

Cancer!

Good jobs came and went with no interest shown by anyone in my résumés and with only little interest in Chris' résumés. We didn't appear to be going anywhere.

In hindsight, it is clear to us now that I began dealing with symptoms of this cancer as early as the summer before I was diagnosed. God knew I had cancer. We did not. He was preparing us for it. We didn't understand.

Chris' job may have been less than wonderful, but her health coverage was excellent. In fact, again in hindsight, it appears that she had some of the best health care coverage offered by any company to anyone who works here in the city of Clemson. Her coverage was better than the state coverage provided to university faculty and staff. We didn't know that at the time, either.

So there we were — continuing along the path — asking God to guide us and show us His will. We were especially praying that He make sure we were in the right jobs. And we were applying for jobs — but nothing was happening.

Other considerations were that Matt, Jon, and Joe were all finishing their respective educations in May 2009. Joe was a senior in high school, Jon was a senior in college, and Matt was a Masters candidate. Those considerations suggested that the 2007/2008 school year was a poor time for us to change jobs or to move. Nevertheless we kept searching and trying.

Then, my rib, chest, and muscle pains began to occur. Initially, they just appeared to be a nuisance — nothing particularly important to be concerned about. But they got worse and worse as the year progressed. Chris became more and more valuable to her boss and the company as the computer person at her location. She also helped her boss guide and direct other co-workers. At one point that winter, every other hourly employee at her location was laid off, but she continued to work. She was needed for the computer work. So again — we continued forward, waiting, watching, and praying for guidance.

Then it hit! Cancer! What a shocker! What were we to do? In hindsight, it appears the Lord had already prepared everything and put wheels into proper motion. He did all this without our understanding any of the situation. He knew! We didn't. He acted! We patiently trusted.

Now we look back and we can see some possible answers. We still aren't 100% certain about these answers, but they at least are beginning to make some sense.

Compliments of Chris' food service job, we are covered by a really good health care plan. Every place I went, when I filled out health insurance forms, they copied my health insurance card and billed the insurance company directly. Other than paying co-pays, we didn't need to do (or pay) anything until after the health insurance paid out. All bills take a month or two for the insurance to process. In October for example, we were finally seeing how the bills from July and August were handled. Our portion was 15%, until we reached our 'total out-of-pocket expense', and then our part dropped to zero.

We didn't know any of this was coming. We were searching for and wanting to change jobs and health insurance companies. But we were working, and God had us covered by a really good plan all that while. We didn't know! He did! Fortunately, we didn't mess it up!

Regarding a support network, we had a huge number of friends and family from all around the country and even around the world praying for us daily. I have many Christian friends at OCA who prayed for us daily. Many students there were also exercised to pray for me, too. When I was a teenager, I didn't look much beyond my own needs for anything. Here was a whole school full of kids who were praying for Dr.

D and family. Every so often, I received notes and cards from the students saying, "We are praying for you!" I really appreciated it! But I also saw that my situation was providing them the opportunity to do their part in active Christian service ... and they appeared to be stepping up to the plate. When visiting personally with them, almost every one said, "We are praying for you!" It is so wonderfully moving to hear that! God certainly knows what He's doing!

Doctors, Clinics, Treatments

How close or how far away are the cancer facilities? As it turns out, every day on the drive from home to OCA, I passed within sight of the local cancer clinic. I could see it each day if I had known where to look. It is within three miles of OCA and exactly along my daily commute. When driving to OCA each day, I neither knew, nor cared, that this was the case. But God knew.

The doctors and staffs at the clinics and hospitals are excellent. Top notch! To have been referred to such a group by our family doctor is wonderful. But our family doctor is a Christian, which means God guided him in his decisions and recommendations as well.

What about my treatment regimen? It appears that my **aggressive** treatment regimen is locally available only through the group of doctors at the cancer clinic. It is available elsewhere, but the Lord put me in just the right spot to be offered the treatment without requiring any further searching or any long trips.

I met a woman with MM from Hendersonville, NC, (1½ hours up the hill into NC from here) who said her doctor was no fan of stem cell transplants until her doctor met my transplant doctor. She had a stem cell transplant with him in Greenville, SC, then, too.

A friend of a friend travels to Arkansas to be treated for her MM. Our treatments are essentially identical, but she has had to travel several hours by plane to Arkansas to be treated while I only needed to drive 30 miles to Greenville, SC.

Another woman treated by my oncologist was diagnosed with MM about a month after my diagnosis. I was told that she has a different health care provider which required her to travel to Florida for her treatments.

The Stem Cell Transplant Director at the clinic in Greenville, who is my transplant doctor, shares an office pod with my primary oncologist. I now know that I was located in the right place at the right time with the best oncologists and the latest treatment techniques without having had to spend any time searching for treatment options. The Lord had it all arranged in advance. Coincidence? I don't think so. The Lord knew the problem I would be facing and He made arrangements for it. He kept me here in the Clemson/Seneca area. He kept Chris in her job with good health care. He made sure we were referred to the proper specialists by our Christian family doctor. Etc.

Need we be worried when the Lord is in control? No. We need to trust that the Lord knows what He is doing. I know this. My struggle is to stay out of His way!

In May, 2009, after all of my chemotherapy was completed, Chris moved to a new office job within the same food service company. She no longer spends hours on her feet in the kitchen, but has a desk job in the front of the conference center building where she interacts regularly with any and all who walk through the door. She doesn't have to work in the kitchen on football weekends any more. The Lord has taken care of her job, too!

My Body's Cooperation with the Meds

Each person's body responds differently to medications and treatment procedures. No one knows ahead of time how he or she will respond. But God knows.

I received many compliments from the medical staff on the way "I" responded to the chemo treatments and medicines. But "I" had nothing to do with it. If my body cooperated fully with the medicines, it was out of my hands. I could strain and concentrate and focus as much as I wanted and that would not have added one iota of strength to the way my body responded to any medication.

I could have sat around all day each day and chanted over and over again, "I will **not** get nauseous from the chemotherapy! I will **not** get nauseous from the chemotherapy! ..." But unless given the proper anti-nausea meds, **and** unless my body cooperated with them, **and** unless they worked properly, chemotherapy **will** cause nausea. God can cause our bodies to respond well and cooperate with anti-nausea meds so that we don't get nauseous. He can do that. We can't. It is out of our control, but it is well within His control!

When "I", meaning my body, cooperated with the treatment program and its meds, "I" actually had nothing to do with it. God gets all the credit!

When "I" received compliments because the medical staff was able to collect the desired number of stem cells from my blood in one day, while other people struggled through several days of collection only to produce a lesser number of cells, "I" had nothing to do with it. God gets all the credit!

When "I" cooperated so well that the concentration of cancer cells in my bone marrow went from 10-15% to less than 0.5% in two months, "I" had nothing to do with it. God gets all the credit!

"I" am simply going along for the ride. I felt the day to day pains and suffered the day to day weaknesses associated with chemo treatments. It was difficult to walk, go up and down stairs, balance, stand up, etc. These were activities I could consciously decide to do (or not to do). I made those decisions. But I couldn't make any decision like, "Today, the bloating in my legs will subside!" That is out of my control. "Today, my blood pressure is going to be perfect!" That, too, is out of my control. "Today, the neuropathy pain in my fingers and toes will be gone." I have no control over that one either. God decides those things.

Fortunately, I know that I am a child of God and His desire is for my best. Therefore, I am in good hands and I need not worry. Whatever He has planned for me will be a wonderful adventure. I just need to be patient as I wait to see where He leads! I just need to stay out of His way and allow Him to reign!

Thank you Lord!

26

My Spiritual Life

When young, I attended a church with my parents. That denomination was and continues to be one of the most liberal protestant denominations. We never carried our Bibles to church and to my knowledge, they didn't have pew Bibles. We were not expected nor encouraged to study our Bibles. The pastor explained all we needed to know on Sunday mornings during **his** sermons. Today, this last sentence would need to be revised. It would need to be, "The pastor explained all we needed to know on Sunday mornings during **his or her** sermons."

To become a Christian in that church, you needed to go through the confirmation process where you had to study a lot of denominational teaching and memorize several scriptural passages. You didn't need to be saved in the sense explained by Romans 3, however. We were taught that God is a loving God who won't cast any good person into hell. As long as you are a part of the **world** (which we all are) and a member of the church (of course), Jesus died for you since He died for the sins of the **world**. They didn't understand the concept of getting saved by faith; they couldn't explain salvation when asked; and therefore salvation (in the Biblical sense) was not a requirement for membership in the church.

To join the church you needed head knowledge about Jesus Christ and church doctrine. Faith, repentance, and a real spiritual conversion was beyond their ability to explain. The real sign that you were a Christian and member of that church was the day you received your own personalized box of collection envelopes in which you were expected to put your weekly offerings before dropping them into the collection plate.

Needless to say, I was taught from small on up that I was a Christian. My folks were. So was I. And that happened without ever having gone through any type of conversion experience. My Mother told me the day of my confirmation would be the most special day of my life, but it passed by like all the rest. Nothing special happened that day. When I inquired further about it, I didn't get a good answer. Neither did confirmation classes provide any clues concerning what to expect that day. It was all a rather empty experience.

Then I went away to college. After a year in New York City, living in an apartment with one of my high school classmates, I transferred to a little school in Upstate New York. The "luck of the draw" put me into a dorm room with a roommate who not only said he was a Christian, but who actually studied the Bible on a daily basis, and acted like one! The group of Christians with whom he congregated were interested in the teachings of the Word of God. That was all new to me.

I learned very quickly that the definition of *Christian* I had been taught as a youngster was very different from their definition. Their definition, however, appeared to be in line with the teachings of the Bible. Mine did not. During the spring semester of 1968, I revised my definition of *Christian* as well as my understanding of what it meant to be a Christian. Then, during the summer of 1968, I accepted the Lord into my life and became a true Christian.

Many awkward moments presented themselves during that transition period, but I survived. For example, my brother and I had memorized several prayers. At dinner time and bedtime, we were always expected to recite those canned prayers. We were taught this from small on up — so we didn't know any differently. We never actually had to put our brains in gear to recite prayers. During that first year at college, I learned that according to the Bible, those prayers were *vain repetitions*. The Lord's prayer out of Matthew 6 was another one of those memorized prayers — another vain repetition — even though it follows a direct admonition in that chapter to "**avoid vain repetitions.**"

The idea of praying with your brain in gear, and actually thinking about what you were saying was a totally new concept to me. Saying, "No," when asked to pray was equally foreign to me. During that time, requests for me to offer thanks for the food at a meal led to awkward moments. It didn't take long for me to learn how to actually think about the words I was praying to God – and that solved the problem.

My true Christian life, therefore, started when I was a 20 year old college student. After considering myself to be a Christian my whole life, during 1967 and 1968, I learned I was not. But after those new revelations, I quickly asked the Lord into my life so I finally and truthfully could say, "Yes, I am a Christian."

Today, I have been a Christian for over 40 years. I do not know everything. I am still learning, and this cancer has opened up new areas of my life and new Christian experiences of which I know very little. I am learning a lot of new stuff at an accelerated pace. I have been learning during all of my Christian life, but the pace seems to have accelerated in recent days due to my new situation.

I have attended local assemblies of Christians since the late 1960s. Christine and I married in 1972 and have participated with various assemblies of Christians ever since. We shopped around a little during the 1990s when our assembly's numbers were down to two families, but we learned very clearly that denominational churches are simply not for us. With denominational churches, the bottom line is the denominational doctrine – not the Bible. When there is any doubt in the meaning of a teaching, the answer must agree with denominational teaching, which doesn't necessarily agree with the Bible. It may. It may not.

Within the assemblies, we have always been taught, and rightly so, that the bottom line is the Word of God. Anything other than the Word of God is not the proper bottom line.

Another problem we learned within denominationalism is that they expect you to 'join their church' before you can actively participate in the local body. Once you sign on the dotted line, you are willingly placing yourself under the *rule* of the local pastor. And unless you agree 100% with his view of the Bible, you will have no opportunity for ministry within that body. In that case, the pastor's understanding of denominational doctrine is the bottom line. That isn't scriptural either.

In the assemblies, we were expected to study and participate in teaching and ministries from square one. No one was expected to sit back and simply warm a pew. Everyone was expected to participate in his or her God-given way. So after years of participation and service in the local ministry, to 'join a church' and then have no opportunity to minister or to serve only with prior approval of the local pastor was foreign to our way of thinking.

This was especially apparent when we heard denominational Sunday school teachers announce, prior to lessons, that what they were about to teach was "approved by the pastor" for use in the class. Then, when they presented things that were non-scriptural, or just plain wrong, the approval of the local pastor meant nothing.

After testing the waters and finding that many of today's organized churches are entertainment centers designed to gather the local community for an hour of all-around entertainment and a little preaching of the Word, we returned to meet with a local assembly of Christians. This meant that instead of meeting with hundreds of others in a large denominational church family, we usually met with only a few believers.

Even then, it is difficult to find believers who are in full agreement. But in small groups, agreement is much more a possibility than in large groups. Many people today do not want to do the heavy lifting required to participate in small groups such as this, or they want to be part of a larger 'church family' (a popular expression here in the South.) One must decide what is important in the Christian life – being entertained on Sundays, being part of a large vibrant church family ruled over by a senior pastor, and only learning a little about the teachings in the Word of God, or participating in the heavy lifting of the study of the Word and its teaching, as well as focusing on a 24/7 spiritual life under the guidance of the Lord.

We have consistently chosen the latter. We have only ever been part of small church families, but we have found, through this cancer ordeal, how many true Christian friends we have made over the years. Families come and go from local meetings, but good Christian friends remain – regardless of where they may be living. Our current support group is not only numerous, but it extends geographically far and wide as a result of our travels and our 40 years of participation in the Lord's church.

Postscript

Having been diagnosed with multiple myeloma, having completed five cycles of chemotherapy including a high dose cycle and an autologous stem cell transplant, and being in the recuperation/maintenance phase of the treatment process, let me say that I would not wish this disease on anyone. The disease is tough! The treatment process is tough! Nothing about this was fun!

As an engineer, I found the equipment, instruments, capabilities of current medical science, and treatment processes quite fascinating! But I would much rather have visited the hospitals and clinics as an engineer on a tour than as a patient who needed these special capabilities and the physicians' expertise.

I wrote this book at the encouragement of friends to help others who may be facing cancer and the treatment process. I have included experiences, explanations, thoughts, feelings, and recommendations. Remember: I am an engineering professor, not a medical doctor — when in doubt about any recommendations, check with your physician and medical staff.

Also, remember: I am a Christian. If family and friends had abandoned me during this process (they did not, but if they had ...), I would not have been alone at any time! The Lord was with me!

If you are diagnosed with cancer, God provides great comfort when you "cast your cares upon Him for He careth for you!" Look to the Lord in all situations! God wants to help everyone. Not everyone, however, wants God's help. For those who don't already know Him, He's waiting for your invitation to help. Invite Him into your life, ask His help, and never be alone!

To all who read this book: God bless you!

www.ingramcontent.com/pod-product-compliance
Lightning Source LLC
Chambersburg PA
CBHW030923180526
45163CB00002B/444